Studies in Systems, Decision and Control

Volume 610

The series "Studies in Systems, Decision and Control" (SSDC) covers both new developments and advances, as well as the state of the art, in the various areas of broadly perceived systems, decision making and control–quickly, up to date and with a high quality. The intent is to cover the theory, applications, and perspectives on the state of the art and future developments relevant to systems, decision making, control, complex processes and related areas, as embedded in the fields of engineering, computer science, physics, economics, social and life sciences, as well as the paradigms and methodologies behind them. The series contains monographs, textbooks, lecture notes and edited volumes in systems, decision making and control spanning the areas of Cyber-Physical Systems, Autonomous Systems, Sensor Networks, Control Systems, Energy Systems, Automotive Systems, Biological Systems, Vehicular Networking and Connected Vehicles, Aerospace Systems, Automation, Manufacturing, Smart Grids, Nonlinear Systems, Power Systems, Robotics, Social Systems, Economic Systems and other. Of particular value to both the contributors and the readership are the short publication timeframe and the world-wide distribution and exposure which enable both a wide and rapid dissemination of research output.

Indexed by SCOPUS, DBLP, WTI Frankfurt eG, zbMATH, SCImago.

All books published in the series are submitted for consideration in Web of Science.

Andrzej Krasuski

Legal Challenges of a Transformation from a Product-Centric to a Service-Centric Business Model and Procedures

Servitisation and Autonomous Vehicles

Springer

Andrzej Krasuski
Faculty of Law and Economics
Jan Długosz University of Częstochowa
Częstochowa, Poland

ISSN 2198-4182 ISSN 2198-4190 (electronic)
Studies in Systems, Decision and Control
ISBN 978-3-031-98900-1 ISBN 978-3-031-98901-8 (eBook)
https://doi.org/10.1007/978-3-031-98901-8

This Springer imprint is published by the registered company Springer Nature Switzerland AG
The registered company address is: Gewerbestrasse 11, 6330 Cham, Switzerland

People say we can survive without manufacturing … that we can have a services economy. But we can't be only about hairdressers and launderettes. An awful lot of services now are linked to manufacturing.

—Feike Sijbesma, CEO of DSM (Milne, 2010)[1]

[1] Milne, R. (2010), Services' wins place at centre of industry, Financial Times, March 24, 2010.

Introduction

This monograph provides a comprehensive analysis of the fundamental legal components relevant to the implementation of digital servitisation within mobility business models, with a particular emphasis on the context of autonomous vehicles, which are distinguished by their complexity and autonomy. The primary research question of the monograph is to assess the legal feasibility of the aforementioned mobility business frameworks.

This investigation examines three fundamental categories of issues.

Firstly, it addresses autonomous vehicles within the context of digital servitisation mobility business models. This collection of issues examines the legal classification of an autonomous vehicle as well as the legal implications of swarm intelligence. This exploration impacts the evaluation of legal relationships established by entities engaged in generating value through the provision of digital mobility services and the underlying technologies, especially concerning autonomous vehicles.

Secondly, it examines the legal status and rights of customers in various roles, including drivers, passengers, and consumers of transportation services provided by autonomous vehicles equipped with digital service options. Distinguishing these aspects is crucial, as customers are the ultimate beneficiaries of the product-service systems linked to the deployment of autonomous vehicles. Given the intricate and multi-stakeholder landscape related to autonomous vehicles, customers face potential risks, particularly concerning privacy protection, cybersecurity, and the effective enforcement of consumer rights.

The third category pertains to the resources essential for implementing digital business models, which encompass digital platforms. Digital platforms play a pivotal role within the ecosystem of entities engaged in creating and delivering digital mobility services. Their market position necessitates establishing appropriate legal regulations to mitigate this influence.

The monograph is founded upon two research theses.

Firstly, regulations constitute one of the pivotal factors that influence the transition of business models from a 'product-centric' to a 'service-centric' paradigm.

Secondly, regulations play a crucial role in determining the acceptability and delivery of digital mobility services utilising autonomous vehicles within a specific ecosystem.

The legal reflection on the research problem delineated and the theses emphasised was conducted using the formal-dogmatic method, commonly known as the linguistic-logical method. This methodology often embraces the research perspective of legal positivism, where the primary aim of the science of law is to describe and systematise legal texts. This method is further complemented by the legal-empirical approach, which fundamentally involves the combined analysis of the content of prevailing laws and the observation of their practical application. The context for implementing artificial intelligence was considered during the selection of relevant provisions. This consideration is evidenced by examples demonstrating the application of an artificial agent in business transactions across various scenarios. It is notable that the scenarios involving the artificial agent were not developed as a result of independent empirical research; rather, they were analysed in relation to factual situations as represented in the existing literature. Perspectives on the issues addressed are presented from international legal literature, alongside pertinent case law. By utilising the aforementioned research sources, an analysis was conducted, facilitating the formulation of de lege lata conclusions and de lege ferenda postulates.

The monograph comprises eight chapters.

Chapter 1 focuses on conceptualising the transformation of business models from product-focused to service-oriented. It explores the causes, objectives, and consequences of this shift towards service-based businesses. The issues raised are examined through the lens of key concepts related to the business model transformation process, acknowledging the complexity involved in this transition.

Chapter 2 addresses the concept and application of autonomous vehicles, providing context for the transformation of business models into service-centric models and the delivery of digital mobility services. Since autonomous vehicles are a sub-domain of artificial intelligence, Chap. 2 explores the concepts of 'artificial intelligence', 'artificial agent', and, based on this, 'autonomous' vehicle. The classification of autonomous vehicles, their operation, and the principles of programming driving algorithms are also examined.

Chapter 3 addresses the significance of law, including socio-technical aspects, in implementing the servitisation of business models. It outlines the relationship between law and the servitisation process while considering the principle of economic freedom. To this end, the concepts of 'law' and 'legal norm' are conceptualised, and reference is made to the process of their creation. Legal risk, particularly that arising from its unpredictability, is also discussed. This chapter presents assumptions regarding the formation of the legal framework for providing digital mobility services. Additionally, sources of law related to the provision of digital mobility services, such as e-commerce services, are introduced.

Chapter 4 addresses the legal aspects of digital mobility services provided within the ecosystem. It proposes a definition of digital mobility services, detailing the categories of services and thoroughly discussing the assumptions made in creating this definition. This definition was developed in the absence of dedicated legislation, considering business models that reflect a shift towards a 'service-centred' approach. These models take into account the legal relationship between the manufacturers of semi- and fully autonomous vehicles and third parties operating within an established ecosystem. The chapter first presents the types of digital mobility services, followed by an analysis of the legal basis for their provision.

Chapter 5 examines the significance of digital platforms in providing digital mobility services. It analyses the complex legal relationships associated with the utilisation of these platforms. Furthermore, the chapter evaluates the market positions of digital platforms and the legislative response aimed at addressing this phenomenon within the economy. Additionally, it discusses the implementation of the Digital Markets Act and the Digital Services Act in relation to the operation of these platforms.

Chapter 6 focuses on the driver of the autonomous vehicle as a recipient of digital mobility services. The driver serves as the link between the digital mobility service and the autonomous vehicle. This chapter discusses the driver's legal status, the requirements for obtaining this status, and the various associated obligations. Determining this status is relevant to the admissibility of digital mobility services. This chapter examines the conditions for granting driver status to an artificial agent operating an autonomous vehicle that moves autonomously and as part of an intelligent swarm.

Chapter 7 examines the sources of privacy regulation and the legal challenges related to the scope, purposes, and legal basis for processing personal data while considering the specific aspects of using autonomous vehicles and providing services through digital platforms. The context of their provision is shaped by autonomous vehicles and involves processing vast amounts of data, including information that enables the identification of individuals. Additionally, the principles of confidentiality regarding electronic communications are examined.

Chapter 8 delves into the legal framework governing cybersecurity. This analysis is conducted through an evaluative lens, scrutinising the extent to which the topic is addressed within EU legislation and identifying the responsible entities. The issue of cybersecurity is significant given the context surrounding the provision of mobility services, particularly in relation to electronic communication methods. Furthermore, it is relevant to the utilisation of autonomous vehicles concerning digital mobility services.

This monograph targets academic researchers and students, as well as legal professionals linked with vehicle manufacturers and their partners in providing digital mobility services. This group includes car-sharing companies, digital platform providers, and technology suppliers supporting these services, such as vehicle connectivity software and positioning technologies.

The monograph takes into account the legal situation as of May 15, 2025.

Warsaw, Poland
May 2025

Associate Professor
Dr. habil. Andrzej Krasuski

Contents

Abbreviations

ADS	Automated Driving System
AI	Artificial Intelligence
AV	Autonomous Vehicle
DTT	Dynamic Driving Task
ENISA	The European Union Agency for Cybersecurity
EV	Electric Vehicle
MAAS	Mobility as a Service
ODD	Operational Driving Domain
OTT Service	Content, a service or an application that is provided to the end user over the public Internet
PSS	Product Service Systems
SAE	Society of Automotive Engineers

List of Tables

Chapter 1
Conceptualising the Transition from a Product-Centric Business Model to a Service-Centric Model

1.1 Introduction

This chapter aims to conceptualise the events in the business of an entrepreneur transitioning from initial business models focused on providing products to those prioritising the provision of services, either independently or alongside products. Understanding this change involves discussing the motivations behind the shift from product-focused approaches to service-oriented strategies. This transformation should be examined through the lens of historical management literature to trace the evolution of thought in this field. Establishing a definitional framework and clarifying the meanings of individual concepts is important for several reasons. First, it designates the research area covered by this monograph. Second, it facilitates clear argumentation in a context where the concepts highlighted in this chapter will serve as subjects or reference points for this discussion. Third, the concepts related to business transformation are not legally defined, and different researchers' perspectives lead to various interpretations of this issue. Therefore, it is crucial to conclude this section with insights into how this transformation is perceived and its characteristics, enabling further analysis in this monograph. The concepts presented in this chapter can be examined through the lens of the goal of transforming original business models.

1.2 Transitioning from a Business Model to a Business Process

At the outset, it is necessary to analyze concepts such as 'business model' and 'business process'. The term 'business model' refers to the logic that organisations follow to create and provide value to the market [62].

A. Krasuski, *Legal Challenges of a Transformation from a Product-Centric to a Service-Centric Business Model and Procedures*, Studies in Systems, Decision and Control 610, https://doi.org/10.1007/978-3-031-98901-8_1

A business model is, therefore, a reflection and tactical realisation of a firm's strategy, but it is not a strategy itself; these two are related but different constructs [14]. A business model is considered a recipe for capturing and defining a firm's behaviour and characteristics concisely [2], and its components and elements illustrate the business logic and how these elements combine to make that business logic work.

A 'business model' is related to the entrepreneur's strategy, facilitating its execution. The term 'business strategy' is best understood as a reflection of organisational choices [52]. Frambach et al. [21] indicate that a firm's strategy is shaped by its cultural values. Furthermore, a 'business model' serves as a tactical framework to articulate how the business strategy will be executed to generate key value propositions- specifically, how to create, deliver, and capture value [69].

According to Cabanelas et al. [12], the basic assumption of the approach is that a firm builds its business model by making various choices to generate revenues encompassing a range of elements such as infrastructure, business offers, customer base, and financial structure. The infrastructure (internal and external organisation, i.e., value chain and network) generates costs but also enables the creation of a value proposition (i.e., an offer) for customers.

Tukker [65] identifies three categories of business models, differentiating them by customer added value.

The first type pertains to business models that rely exclusively on the product. Here, the product itself holds value; for example, possessing or renting a car is valuable to the customer.

The second type pertains to purely service-based business models, where the content of the service serves as the basis for value creation. Examples include the provision of motor insurance or car repairs.

The third type is based on the functions of business models, covering three typologies. For the transport system, these categories are as follows:

(1) A product-oriented system where the user owns the product, but the vendor adds value through advice on efficient use (perhaps via training courses) or maintenance services. The customer is responsible for maintenance and end-of-life management.
(2) A use-oriented system, where the customer utilises a service, such as a bus journey, and the vendor is responsible for the quality of the outcome (e.g., choice of route, driver, or vehicle type).
(3) An outcome-oriented system, where the customer pays for the mobility service, but the service provider determines the best transport option (e.g., route plan) and the most suitable infrastructure option (e.g., road vehicle).

In this analysis, as articulated by Vargo et al. [68], the concept of 'service' is described as the process through which a service provider utilises its resources for the advantage of a customer. Conversely, a 'business process' consists of a series of activities that convert various inputs into outputs that provide value for the customer. Each business process requires a clearly defined objective, which is influenced by external events. Furthermore, additional processes affect the realisation of this objective. Lindsay et al. [38] assert that sustainable business processes executed by human operators necessitate a delicate balance between learning from past experiences and preparing for the future, while reconciling regulations and constraints with the need for freedom and flexibility.

A business model is distinguished from a business process in the following manner. A business model functions as a strategic framework for business strategies, while business processes are implemented at the operational level as methods for actualising that business model within an organisation's structure, systems, functions, and associated workflow mechanisms.

Further analysis should integrate key elements of the business model approach: 'a business model is a conceptual tool that contains a set of elements and their relationships and allows expressing a company's logic of earning money' [48].

1.3 Transformation of Business Models from Product-Centric to Service-Centric

The initial phenomenon of transitioning business models from a product-centric approach to a service-centric one was acknowledged in management literature by Vandermerwe et al. [67]. In examining the transformation process, the term 'servitisation' (or 'servitisation of business') was initially introduced. This term pertains to a business model wherein firms provide 'bundles' composed of customer-focused combinations of goods, services, support, self-service, and knowledge, with the stipulation that within this mix, services assume the most significant role. It is essential to emphasise that the strategy pertains to the transformation of both the original product-centric business models and activities that were inherently service-oriented. Consequently, the strategy proposed by Vandermerwe et al. [67] also involves the transformation of the original service business. When contemplating transformation, one should consider the original outcome of this activity, manifested in the form of a specific product or service. The authors adjust the sequence of the modules within this bundle, contingent upon whether the foundational aspect was the production of goods or the delivery of services.

Initially, Vandermerwe et al. [67] delineate the following sequence of elements: product, service, knowledge, support, and self-service. In the next instance, the order is information, service, support, product, self-service, and knowledge. Each module is capable of functioning independently, demonstrating its free-standing nature. Furthermore, the aforementioned sequence is dynamic, as it is oriented towards the consumer and should be tailored to the specific needs of the customer in a given situation. Consequently, the proposed strategy emphasises the provision of customised solutions for customers, considering the available array of modules that comprise the bundle. Servitisation, in its original context, should be interpreted in a broader sense than merely delivering maintenance services associated with a specific product, such as those pertaining to warranty obligations.

Baines et al. [4] propose that 'the innovation of a manufacturing organisation's capabilities and processes to shift from selling product to selling an integrated product and service offering that delivers value in use' constitutes the process of servitisation. Tukker [65] delineated various forms of this process along a product-service continuum, extending from products augmented by services as an 'add-on' to services supplemented by tangible goods as an 'add-on'.

In this regard, Kowalkowski et al. [33] assert that the notion of service transition, established in the path-defining study by Oliva and Kallenberg [47], posits that firms engage in a unidirectional repositioning along the product-service continuum: progressing from basic, product-oriented services toward more customised, process-oriented offerings, which ultimately culminates in solution delivery. Furthermore, Oliva et al., in their position paper, introduced the concept of unstructured servitisation.

The literature also indicates a transition from products to solutions through the advancement of services. Specifically, this transition involves moving from less sophisticated services to more sophisticated services and, ultimately, solutions along a product-service continuum. Empirically, this transition is frequently described as 'service infusion' [49].

The evolution of business models towards acquiring solutions, without categorising this process as servitisation, has also been noted in the perspectives of various researchers [6, 20, 59, 71]. In the viewpoints held by scholars regarding the transitionship effect, the concept of 'integrated solutions' [11] is similarly referenced. This concept should also encompass modules such as products, systems, and services that are tailored to meet customer requirements and fulfil their needs. However, the type and quantity of these modules differ from those presented in the original concept [67]. The term 'system' as utilised in the aforementioned concept [11] is defined by the authors as a collective term for hardware + software + systems. Ultimately, the concept of 'system' encompasses the same idea, yet it remains inadequately explained.

In alternative perspectives presented in the literature [19], the provision of a customised solution is not regarded as the objective of the transformation process but rather as an alternative to a singular service. This distinction holds significance, as within the previously articulated transformation concepts, service constituted one of the solution modules. Concurrently, Fang et al. [19] emphasised that manufacturing firms ought to transition towards 'solution' and/or 'service' offerings in order to enhance their competitive positioning amidst an era characterised by fierce global competition and escalating commoditisation prevalent in numerous product markets, referencing the literature on the topic [58, 68]. The alternative aims of transformation identified by Fang et al. [19] are termed strategic redirections, which are explicitly classified as 'service transition strategies'.

Other researchers from the initial period of analysis regarding the transformation issue [17] also fail to designate this process as 'servitisation'; instead, they refer to it as 'service management', which represents a managerial philosophy not exclusive to service firms. The 'service management' delineated by the authors encompasses the following components: service package + customer involvement + open access logic. The service package pertains to the collection of activities and information/operational interactions requisite for establishing the relationship with the customer and facilitating delivery. Consequently, the delivery transaction initiates a series of services (in terms of research and development, design, logistics, etc.) characterised by both tangible and intangible elements. Customer involvement in the goods delivery process positions the customer as both a 'buyer', since they receive the final object of the transaction, and a 'supplier', as they benefit from a variety of services offered by the intermediary at the lower end of the chain aimed at ensuring the comprehensive and positive realisation of the object itself. Furthermore, the open system logic may be implemented where enhancements are required in the design, production, and logistics sectors (including, among other factors, information exchange, management transparency, integrated perspective, and delivery management).

Baines et al. [3] presented a different notion from Vandermerwe et al.'s [67] concept of servitisation. The authors define the new concept as an 'innovation of an organisation's capabilities and processes to better create mutual value through a shift from selling product to selling Product Service Systems (PSS)'. The PSS framework aims to facilitate the 'bundling' of products and services.

This should be viewed as a connection to the concept presented by Vandermerwe et al. [67], which signifies the means to implement servicisation. Goedkoop et al. [24] initially defined Product-Service Systems (PSS) as a system comprising products, services, networks of 'players', and supporting infrastructure designed to satisfy customer needs while simultaneously reducing the impact on the environment compared to traditional business models [7]. Goedkoop et al. [24] originally characterised a PSS as a system encompassing products, services, a network of 'players', and a supportive infrastructure tailored to fulfil customer requirements while concurrently minimising environmental impact relative to conventional business models.

Furthermore, Neely [45] articulate a perspective on the establishment of a Product-Service System (PSS) as a target of transformation, understood as an integrated offering of products and services that provide value in use. These authors further emphasise that the concept of 'Global Value System' relates to the worldwide network of suppliers, customers, and partners who must collaborate to ensure that integrated product and service offerings yield value in use. Moreover, the modification of business models that increasingly emphasise services may also be driven by technological advancements, particularly through Information and Communication Technology (ICT) or cloud-based systems [7]. Subsequently, Chesbrough [75] introduced the idea of 'open service innovation', perceived as the enhancement of an organisation's innovation capabilities through a transition from products to product-service systems, thereby more effectively addressing customer needs and avoiding the trap of commoditisation.

While the emphasis on the aforementioned concepts of servitisation has predominantly centred on the object aspect of these transformations, the existing literature has also acknowledged the subjective aspect of servitisation [29]. This subjective aspect must be evaluated in two dimensions. Firstly, within the framework of industrial networks and business-to-business relationships, the significance of product-service transitions is pertinent to the readership of Industrial Marketing Management. Secondly, it is essential to consider the relationship between the manufacturer and the customer.

In reference to the preliminary analysis of business-to-business relationships, Gadde et al. [22] contend that in an industrial network, the characteristics of prospective interdependence and coevolution hold significant importance, while the competitive aspect of strategy diminishes in relevance. Furthermore, within the domain of relational strategy, recent developments reinforce the idea that the core of strategy resides in the ability to establish and nurture relationships with other entities. Interdependent processes (e.g., maintenance activities that are integrated within business networks) significantly affect the performance, productivity, value creation, and revenues of various entities, thereby necessitating an interorganisational framework for their analysis [26].

Servitisation is also delineated within the framework of intra-organisational modifications that arise from decisions made to adopt a 'service-based business model' [30]. The pertinence of these modifications to servitisation lies in their intended purpose to facilitate its execution [31]. These modifications are designed to foster collaboration between users and producers, promoting a trajectory towards long-term partnerships [29]. Consequently, a customer-centric approach is articulated. The execution of servitisation necessitates comprehensive adjustments within the organisation [63], for which robust organisational leadership (i.e., service leadership capability) is essential [31].

In the second entity approach, it must be emphasised that servitisation cultivates an understanding and appreciation of value-in-use for customers [29]. The concept of 'value-in-use' has been defined in the literature [40] as the outcome, purpose, or objective achieved by a customer through service. According to the authors, the term 'value-in-use' establishes a significant connection between service quality, on one hand, and relationship outcomes, on the other.

1.4 Definition of Servitisation

Servitisation pertains to the relationship between the manufacturer and the customer from an external perspective. Moreover, it encompasses an internal dimension, focusing on intra-organisational transformations.

According to Vandermerwe et al. [67], 'servitisation' is occurring across nearly all industries worldwide. Influenced by the dynamics of deregulation, technology, competitive pressure, globalisation, and significant competition from both service companies and manufacturers, this trend is evident in various sectors.

The concept of 'servitisation' must be understood within the frameworks of theoretical constructs, empirical phenomena, and research domains. As the subsequent analysis detailed in the book pertains to empirical examination, the empirical methodology will inform the development of this definition.

The concept of 'servitisation' should be related to the economic activities of a particular entrepreneur. Therefore, it must be examined both internally and externally, using the entrepreneur's sphere of influence in their business activities as a basis for differentiation.

A characteristic of servitisation is the transformation from the initial state to a revised state. In the initial state, the manufacturer is engaged in producing goods for sale or supply (servitisation may also pertain to service providers). This initial state undergoes modification through a reassessment of the manufacturer's mission, vision, and business strategy, or more accurately, the business model, transitioning from one that was originally focused on providing products to one that is centred on delivering services exclusively or offering services alongside products or other functionalities, depending on the customer's requirements. The existing literature occasionally describes this as the provision of customer solutions [61].

For contemporary manufacturers to deliver goods and services proficiently—either independently or in conjunction with a product or additional features—implementing intra-organisational changes is imperative. This transformation necessitates a novel internal perspective on 'Servitisation'. It entails intra-organisational modifications, including reconfiguring organisational structures and altering resource acquisition and utilisation methods. The magnitude of these changes within the entrepreneurial organisational domain, along with resource management, is justified by the transition from a product-centric to a service-oriented business model. An internal strategy may also require establishing legal relationships with third parties to facilitate the reorganisation of structures and resources. Fundamentally, servitisation encompasses the intra-organisational activities undertaken by the entrepreneur, which enable the provision of services to customers or services associated with goods, as part of a business model grounded in a service-centric methodology. According to the literature [32], servitisation demands a reassessment of business logic, which encompasses the firm's raison d'être and the mental models of managers. While the concepts of business logic and business models are frequently perceived as interconnected, numerous interpretations of the business model neglect these 'softer' elements (Wirtz et al. 2016). This dimension must be addressed internally.

The external perspective on 'servitisation' pertains to the application of the service business model within customer relationships. According to Kowalkowski et al. [32], a service business model encompasses the supplier's commitment to enhancing the value that customers derive from usage, thereby assuming greater responsibility for the entire value-creation process compared to traditional product-focused, transaction-based models. Consequently, all activities essential for delivering a service—whether standalone, alongside a product, or integrated with other functionalities or solutions—should be included. As a result, the external perspective embodies the objective of servitisation by implementing the service business model to enable customers to realise this value in use. Literature, including Kowalkowski et al. [32], emphasises this increased responsibility for value creation in relation to product-centric, transaction-based models. Within this context, the paradigm of service-centricity is viewed as synonymous with customer-centricity.

The servitisation literature (e.g., [15]) identifies several opportunities for manufacturers associated with a service-centric approach to business models. This approach is regarded as a strategic alternative to product innovation [13] and standardisation [4], serving as a means to address commoditisation [41] and as a method to cultivate unique, loyal customer relationships [65]. For customers, servitisation has the potential to enhance value through more customised and integrated product offerings that effectively meet their needs [4, 8].

The transition to a service-oriented business model is expected to yield profits for the entrepreneur. Boulding et al. [9] argue that the fundamental principle of the customer-centricity paradigm is not focused on the process of selling products but rather on creating value for the customer, which also fosters value for the firm; in essence, customer centricity involves the methodology of dual value creation. The literature [32] indicates that the revenue mechanism of the service model depends on the outputs derived from customer value-creating processes, such as ensuring a specified level of product availability or achieving a predetermined level of performance, rather than relying on inputs like the number of service hours sold. If the implementation of servitisation fails to produce financial gains, a servitisation paradox arises, as elaborated in Sect. 1.8.

The concept of 'servitisation' should not be confined solely to producers of goods [67]. For instance, service-oriented enterprises, such as banks, which adhere to a product-centric logic (seeking to maximise the sales of finished 'financial products' while simultaneously distancing themselves from their clientele through automation and digitisation), may embrace a product-focused mindset and business rationale [60]. Similarly, as outlined by Grönroos [25], a supplier of consumer durables or industrial products has the capacity to adopt a service-oriented logic that emphasises not the products themselves but the processes through which these products are integrated, thereby enabling the emergence of customer value. In summary, a company predominantly based on services, with a notably significant beneficial owner, may pursue a product-focused logic and conversely [32].

1.5 Services Integrated Within Products

Regarding transportation, it is essential to emphasise that transitioning to service-centric business models is necessary, as vehicles (considered as products) embody services that contribute to value creation for their manufacturers and related partners. Certain products inherently integrate services. Examples of such products include, but are not limited to, photocopiers and vehicles.

Lightfoot et al. [3] presented illustrations such as Xerox's transition from selling printers and copiers to providing a 'Document Management Service' [72],

Rolls-Royce Civil Aerospace's 'TotalCare' Service, and Alstom's engagement in supporting the UK west coast mainline for Virgin Rail through its Train Life Services.

Sawhney [58] presented an instance of how Kodak conceptualised its digital imaging operations in relation to the customer activities involved in managing memories. The cycle of client activity commences with the 'capture of memories' using a camera, camcorder, or camera phone. Subsequently, these memories undergo modification and editing, facilitated by digital image manipulation software. The images can then be created, stored, transmitted, shared, archived, and repurposed. Based on this cycle, Kodak has developed a diverse range of services, including online photo printing services, paper and printers for home photo printing, as well as card and flyer printing services using photographs. Therefore, Kodak has redefined its identity as a company that assists customers in managing their memories rather than merely producing cameras and film. The company has successfully produced compelling solutions such as the EasyShare camera system, photo printing services, and Picture Maker kiosks located in pharmacies for the self-printing of digital photographs.

It is emphasised in the literature (e.g., [67]) that this 'materialisation', while not a new phenomenon, is becoming increasingly prevalent and significant. This shift is altering the nature of the service, its delivery, and the potential for internationalisation. The key point, however, is that a larger component of the value added to customer offerings is being allocated to services. Servitisation represents a complex process that transcends the mere 'servicing' of products. It necessitates strategic decisions at the highest levels of management. Such decisions may impact the development of customer relationships and, consequently, the creation of value, contributing to revenue generation and the advancement of the entrepreneur.

The trend towards increased interest in product functionalities is also reflected in government administration. According to Neely [45], governments worldwide assert a preference for contracting capabilities rather than purchasing specific products. For instance, the United Kingdom's defence industrial strategy specifies that the Ministry of Defence seeks to procure and execute operations, rather than merely acquiring the physical equipment. Consequently, there has been an increase in outsourced support services provided by firms such as BAE Systems and Rolls-Royce.

Another example is Volvo, a manufacturer of passenger cars, trucks, and buses. It recognised early on the benefits of value creation arising from its customers' automotive-related activities, including finance and insurance, petrol stations, and roadside assistance networks that consist of towing and various emergency systems [67].

1.6 The Influence of Digitalisation on the Development of Service-Centric Business Models Within the Transportation Sector

The shift from a product-oriented to a service-oriented business model is driven by digitalisation [1]. Technologies associated with digitalisation are reshaping the automotive industry and fostering new opportunities and business models [39], emphasising value creation. Business models concerning digital mobility are founded on the transformation of processes and procedures necessitated by digitalisation, as well as the imperative to integrate knowledge and technology to generate customer value [73].

The term 'digitalisation' lacks a legal definition; however, it has been articulated in academic contexts. Hsu [28] characterised 'digitalisation' as the integration of digital technology to facilitate connections among individuals, systems, corporations, products, and services. Essentially, 'digitalisation' denotes tools designed to enable the exchange of information between entities and objects, with the ultimate objective of benefiting users. From an economic standpoint, these information exchange tools should not only fulfil their initial purpose effectively but also confer advantages—especially compared to analogue communication—by enhancing operational performance and expanding the capabilities of corporations [43].

According to Coreynen et al. [15], the following stages exist for using the term 'digitalisation'. These stages encompass production, marketing, sales, and customer relationship building.

During the initial phase, specifically the production stage, digitalisation assists manufacturers in implementing enhancements to back-office operations. These enhancements include:

(1) the effective and efficient creation of solutions throughout the goods production process, such as the digitisation of standardised solution modules into the firm's enterprise resource planning or product data management systems [61];
(2) improvements in operational efficiency, for instance, through the automation of production processes [42], and
(3) an increase in transparency, which encompasses enhanced monitoring and documentation of both physical and functional states of products throughout the stages of production, construction, usage, and recovery [46].

According to Coreynen et al. [15], a manufacturer can leverage this knowledge to enhance customers' processes by offering advisory or training services. This facilitates the manufacturer's integration into the customer's operational processes [65].

Secondly, from a commercial or 'front-end' perspective, providers strive to acquire deeper insights into the value-creating processes of customers [50] and facilitate the achievement of their objectives [8]. Front-end digitalisation exerts a more immediate influence on the customer in comparison to back-end digitalisation, which should be implemented initially. This is evidenced by the sequence whereby companies first adopt digital management systems, followed by technologies facilitating wireless human–machine communication, and ultimately, technologies associated with cyber-physical systems [74].

According to Coreynen et al. [15], the third pathway involves the creation of digitally-enabled offerings that fundamentally alter customer processes and enhance provider-customer relations. This growing digitalisation in mobility is bolstered and directed by sustainability, connectivity, and autonomous vehicles [5].

Sustainability pertains to the implementation of electric vehicles (EVS) and the mitigation of emissions [35]. Various service propositions based on digitalisation include, for instance, digital services associated with charging stations, where the navigation system can identify the most suitable charging station [51]. Moreover, an optimised navigation system has the potential to enhance the efficacy of electric car rental systems by minimising average travel time and reducing energy consumption [56].

Connectivity refers to the transmission of various types of data exchanged among vehicles, individuals, and mobility infrastructure, as well as facilitating access to and sharing of this data [34]. The potential contribution to environmental sustainability through car-sharing primarily lies in its ability to reduce both vehicle production and the number of kilometres driven [64]. Consequently, a correlation exists between the domains of 'connectivity' and 'sustainability'. Therefore, connectivity is an indispensable tool for providing digital mobility services. Moreover, personalised, connected services and enhanced user experiences depend on digitalisation, exemplified by better vehicle usage (e.g., vehicle status information, remote servicing) or entertainment (e.g., differentiated content or access to social media), along with software updates selected by the customer and typically funded through subscriptions [35].

A correlation exists between digitalisation and autonomous vehicles. Digitalisation exerts a direct influence on the utilisation of autonomous vehicles, and the associated business models are significantly transforming the market [5]. This evolution enables automakers to develop various digital business models pertinent to autonomous vehicles [57]. Specifically, business models can encompass diverse aspects of vehicle usage, including autonomous vehicles, fleet services [36], and advanced driving solutions [60].

1.7 Rationale for the Implementation of Servitisation Business Models

In discussing the underlying reasons for servitisation, it is prudent to reference the perspective of Oliva et al. [47], who argued that management researchers have advised product manufacturers on the necessity of integrating services into their product offerings. This is supported by works such as Bowen et al. [10], Gadiesh et al. [23], Quinn et al. [54], and Wise et al. [71], which highlight the higher profit margins associated with services. Furthermore, another rationale for servitisation is the establishment of capabilities for differentiation through supplier and customer relationships [55]. The predominant intent behind servitisation strategies appears to enhance competitive positioning by concentrating on service provision, as stated by Baines et al. [3]. According to Harmon et al. (2012), services are characterised by diminished visibility and are inherently based on relationships. These characteristics facilitate value creation, making the services difficult to replicate [19]. Additionally, another aspect emphasised in the literature is that the motivation for engaging in servitisation originates from the pursuit of competitive advantage through unique, customised services [31].

Environmental protection can serve as a catalyst for servitisation. It is feasible to mitigate the adverse environmental impact of products if companies alter their business models and customers reevaluate their notions of ownership [24]. In this regard, Neely [45] cite the example of rented washing machines. Customers are increasingly less interested in owning a washing machine, rather, they prioritise the functions of such a machine, specifically the outcomes of the washing process. Consequently, customers exhibit a willingness to pay a fixed fee per washing cycle. In a service-oriented business model that emphasises enhanced customer relationships, it is in the customer's interest to minimise the number of washes, thereby reducing costs. Similarly, it is in the provider's interest to extend the product lifecycle [44]. Once the machine is installed, the provider's interest in maintenance work diminishes.

The aforementioned concept is based on the assumption that customers expect more specific, customised services, thereby increasing the need for specialisation. Furthermore, services represent a more stable source of revenue, as they show resilience to the economic cycles that affect investment and equipment acquisitions [53]. Oliva and Kallenberg [47] also assert that there is a competitive rationale. Services, because of their lower visibility and greater labour-dependence, are significantly more difficult to replicate, thus establishing a sustainable source of competitive advantage [27].

Oliva and Kallenberg [47] accurately acknowledge that the transition to service-centric business models is linked to the life cycle of products. They contend that durable manufactured products, such as capital equipment and consumer durable goods, require services as they progress through various stages of their life cycle, including acquisition, installation, operation, upgrades, and decommissioning. Furthermore, they highlight the cost of ownership associated with these products, which extends beyond the initial purchase price, encompassing expenses related to spare parts, consumables, and maintenance.

The concurrent delivery of services alongside the manufacturing of goods necessitates an examination of the interplay between service and product quality. Enhancing the quality and variety of services extends product longevity and reduces the need for new products. Conversely, boosting the quality and durability of products can potentially decrease future service revenues.

Nonetheless, the provision of services is linked to significant advantages for the manufacturer, including customer loyalty, increased pricing power, and enhanced resilience against outsourcing.

Upon examining the issue from the perspective of service characteristics, the ensuing benefits associated with its provision become evident. Services rendered alongside delivered products enhance the company's proposition, introducing a level of 'intangibility' that complicates efforts to anchor it within standardisation processes. Nevertheless, strategies designed for service transformation may make a company's overall offering, comprising both tangible and intangible value propositions, more distinctive, challenging to replicate or imitate and more tailored to address customer needs. These characteristics are expected to bolster the company's value by fostering increased and more stable sales, cash flow, and profits.

Service transition strategies may influence firm value through the following mechanisms: leveraging knowledge and resources, enhancing customer loyalty, potential loss of strategic focus, and organisational conflict [19].

1.8 Deservitisation

The concept of 'deservitiziation' encompasses various interpretations within the existing literature. According to Kowalkowski et al. [32], it is defined as a process in which an entrepreneur withdraws from specific service initiatives. However, this definition is not universally accepted within scholarly discourse. Valtakoski [66] highlighted two distinct perspectives on this concept. The first perspective asserts that entrepreneurs either revert to implementing a product-centric business model or completely abandon the service sector, ceasing the provision of related business activities. Conversely, the second perspective posits a notion of 'reverse servitisation,' wherein entrepreneurs transition to offering customer-oriented services that necessitate 'customer intimacy.' Subsequently, they establish their enterprises based on product-oriented service business models [16].

Deservitisation necessitates managerial decisions that can be executed when a servitisation failure occurs. Consequently, a servitisation failure may arise from implementation deficiencies.

A 'servitisation failure' refers to an instance in which an entrepreneur has failed to cultivate a profitable service-centric business model, potentially leading to a decline in both revenue and customer trust. This condition is referred to in scholarly discourse as a 'servitisation paradox' [70].

The phenomenon termed the 'servitisation paradox in manufacturing companies' was initially defined by Gebauer et al. [32]. Neely [45] highlighted the risks associated with transitioning from a product-centric to a service-centric business model. According to Li et al. [37], the implementation performance of servitisation constitutes a complex endeavour, as it necessitates transformation across various domains both within and outside the organisation, including organisational structure, corporate culture, resource allocation, operational modes, and relationships with internal and external stakeholders. The authors also reference 'uncontrollable' external factors. While these factors encompass the overall industrial environment as well as the global economic or financial environment, they may also pertain to the legal framework.

1.9 Conclusions

The conceptualisation of transforming an entrepreneur's business from a product-oriented to a service-oriented business model is intricate. It requires a preliminary elucidation of a collection of concepts that facilitate understanding of the importance of enterprise transformation. This analysis is justified by the lack of a legal definition regarding business model transformation. Moreover, the discussion of the topics addressed in this chapter arises from the need to standardise the interpretation of concepts present in scholarly discourse that relate, either directly or indirectly, to the process of business transformation towards a model based on creating value through services.

The introductory concept that needs to be addressed in this chapter's scope is the 'business model'. This concept reflects the fundamental purpose of business activity, namely, creating market value. The business model serves as a reflection and tactical implementation of the company's strategy; it is not the strategy itself. Although related, the two are distinct constructs. Additionally, it communicates how to implement key business values, thereby reflecting the entrepreneur's strategy, including their mission and vision. By utilising the business model, the entrepreneur can answer the question of how to create, deliver, and capture value. It specifically addresses key elements such as infrastructure, business offering, customer base, and financial structure.

Value creation can also be examined from the perspectives of both the entrepreneur, serving as the supplier of products and services, and the customer, acting as the recipient. This analysis reveals mutual benefits. Regarding transportation, the following relationships between entities can be highlighted: (1) the user owns the product, while the supplier adds value by providing guidance on its effective use; (2) the customer interacts with the service, and the supplier is responsible for the quality of the outcome (for example, the choice of route, driver, or type of vehicle); (3) the customer pays for the mobile service, while the service provider selects the optimal transport option (for instance, the route plan) and the most appropriate infrastructure choice (such as the road vehicle).

Value creation may also be analysed from the perspective of both the entrepreneur, acting as a supplier of products and services, and the customer, serving as the recipient. This dynamic indicates the presence of mutual benefits. In the context of transportation, the following relationships between entities can be established: (1) the user possesses ownership of the product, while the supplier enhances its value by offering guidance on effective utilisation; (2) the customer engages with the service, and the supplier assumes responsibility for the quality of the outcome (e.g., choice of route, driver, or type of vehicle); (3) the customer remits payment for the mobile service, yet the service provider is tasked with determining the optimal transport option (e.g., route plan) and the most suitable infrastructure option (e.g., road vehicle).

The transition from products also includes the acquisition of solutions through the development of services. This indicates that the transition moves from less advanced services to progressively more advanced ones, culminating in solutions that cover the entire product-service continuum. Empirically, this transition is often called 'service infusion'. The aforementioned considerations require the clarification of the concept of 'servitisation', which should be understood within a theoretical framework, as an empirical phenomenon, and as a research area. This concept relates to the relationship between the manufacturer and the customer, necessitating intra-organisational changes. Such changes involve the mission, vision, and business strategy of the manufacturer; specifically, the transformation of the business model from one that initially prioritised product provision to one that emphasises either the exclusive offering of services or the simultaneous provision of services alongside products or other functionalities, depending on customer needs. The extent of the changes introduced within the entrepreneur's organisational sphere, as well as resource management, is warranted by the shift of the business model from a originally product-oriented approach to a service-oriented one.

When transforming business models, entrepreneurs should also consider access to digital tools that facilitate communication with customers and support data analysis for service provision. Functionally, the definition of 'digitalisation' refers to tools for information exchange between entities and things. In the latter case, exchanging information between things aims to serve subjects. In relation to transport, it should be emphasised that the transformation towards service-oriented business models is justified because vehicles (as products) include services that create value for their manufacturers and partners. Some products have services embedded in them, such as photocopiers and vehicles. A larger component of added value in customer offers is attributed to services. Servitisation is a complex process that extends beyond simple 'servicing' of products. It requires strategic decisions at the highest level of management. This decision can impact customer relationship building and, consequently, value generation, contributing to revenue growth and the development of the entrepreneur. In servitisation terminology, it is also necessary to distinguish the reverse process called 'deservitisation', which arises from the failure of the applied business models. This term encompasses two situations. On one hand, 'deservitisation' should be understood as a return to 'product-centric' business models. The second solution involves discontinuing the business activity.

References

1. Ardolino, M., Rapaccini, M., Saccani, N., Gaiardelli, P., Crespi, G., Ruggeri, C.: The role of digital technologies for the service transformation of industrial companies. Int. J. Prod. Res. **56**(6), 2116–2132 (2018)
2. Baden-Fuller, C., Morgan, M.S.: Business models as models. Long Range Plan. **43**(2), 156–171 (2010). https://doi.org/10.1016/j.lrp.2010.02.005
3. Baines, T.S., Lightfoot, H.W., Benedettini, O., Kay, J.M.: The servitisation of manufacturing: a review of literature and reflection on future challenges. J. Manuf. Technol. Manag. **20**(5), 547–567 (2009). https://doi.org/10.1108/17410380910960984

4. Baines, T., Lightfoot, H.W., Evans, S., Neely, A., Greenough, R., Peppard, J., Wilson, H.: State-of-the-art in product-service systems. Proc. Inst. Mech. Eng. Part B: J. Eng. Manuf. **221**(10), 1543–1552 (2007). https://doi.org/10.1243/09544054JEM858
5. Barreto, L., Amaral, A., Baltazar, S.: Mobility in the era of digitalisation: thinking mobility as a service (MaaS). In: Goncalves, R., Sgurev, V., Jotsov, V., Kacprzyk, J. (eds.) Intelligent Systems: Theory, Research, and Innovation in Applications. Studies in Computational Intelligence, pp. 275–293. Springer, Cham (2020)
6. Bennett, J., Sharma, D., Tipping, A.: Customer solutions: building a strategically aligned business model. Insights: Organization and Strategic Leadership Practice, pp. 1–5. Booz Allen & Hamilton (2001)
7. Berman, S.J., Kesterson-Townes, L., Marshall, A., Srivathsa, R.: How cloud computing enables process and business model innovation. Strategy Leadersh. (2012)
8. Bettencourt, L.A., Brown, S.W.: From goods to great: service innovation in a product-dominant firm. Bus. Horizons **56**(3), 277–283 (2013). https://doi.org/10.1016/j.bushor.2013.01.008
9. Boulding, W., Ajay, K., Richard, S., Zeithaml, V.A.: A dynamic process model of service quality: from expectations to behavioural intentions. J. Mark. Res. **30**(1), 7 (1993)
10. Bowen, D., Siehl, C., Schneider, B.: Developing service-oriented manufacturing. In: Kilmann, I. (ed.) Making Organizations Competitive, pp. 397–418. Jossey-Bass, San Francisco, CA (1991)
11. Brady, T., Davis, A., Gann, D.M.: Creating value by delivering integrated solutions. Int. J. Project Manag. **23**, 360–365 (2005). https://doi.org/10.1016/j.ijproman.2005.01.001
12. Cabanelas, P., Parkhurst, G., Thomopoulos, N., Lampon, J.F.: A dynamic capability evaluation of emerging business models for new mobility. Res. Transp. Bus. Manag. **47**, 100964 (2023). https://doi.org/10.1016/j.rtbm.2023.100964
13. Carlborg, P., Kindström, D., Kowalkowski, C.: The evolution of service innovation research: a critical review and synthesis. Serv. Ind. J. **34**(5), 373–398 (2013). https://doi.org/10.1080/02642069.2013.780044
14. Casadesus-Masanell, R., Ricart, J.E.: From strategy to business models and on to tactics. Long Range Plan. **43**(2), 195–215 (2010). https://doi.org/10.1016/j.lrp.2010.01.004
15. Coreynen, W., Matthyssens, P., Van Bockhaven, W.: Boosting servitization through digitization: pathways and dynamic resource configurations for manufacturers. Ind. Mark. Manag. **60**, 42–53 (2017). https://doi.org/10.1016/j.indmarman.2016.04.012
16. Cusumano, M.A., Kahl, S.J., Suarez, F.F.: Services, industry evolution, and the competitive strategies of product firms. Strateg. Manag. J. **36**(4), 559–575 (2015)
17. DeToni, A., Nassimbeni, G., Tonchia, S.: Service dimensions in the buyer-supplier relationship: a case study. Int. J. Phys. Distrib. Logist. Manag. **24**(8), 4–14 (1994)
18. Eight types of product–service system: eight ways to sustainability? Experiences from SusProNet. Bus. Strateg. Environ. **13**(4), 246–260. https://doi.org/10.1002/bse.414
19. Fang, E., Palmatier, R.W., Steenkamp, J.-B.E.M.: Effect of service transition strategies on firm value. J. Mark. **72**(4), 1–14 (2008)
20. Foote, N., Jay, G., Hope, Q., Miller, D.: Making solutions the answer. McKinsey Q. **3**, 84–93 (2001)
21. Frambach, R.T., Prabhu, J., Verhallen, T.M.M.: The influence of business strategy on new product activity: the role of market orientation. Int. J. Res. Mark. **20**(4), 377–397 (2003). https://doi.org/10.1016/j.ijresmar.2003.03.003
22. Gadde, L.-E., Huemer, L., Håkansson, H.: Strategizing in industrial networks. Ind. Mark. Manag. **32**, 357–364 (2003)
23. Gadiesh, O., Gilbert, J.L.: Profit pools: a fresh look at strategy. Harv. Bus. Rev. **76**(3), 139–147 (1998)
24. Goedkoop, M., van Halen, C., te Riele, H., Rommens, P.: Product service-systems, ecological and economic basics. Report for Dutch Ministries of Environment (VROM) and Economic Affairs (EZ) (1999)

25. Grönroos, C.: Adopting a service logic for marketing. Mark. Theory **6**(3), 317–333 (2006)
26. Hedvall, K., Dubois, A., Lind, F.: Analysing an activity in context: a case study of the conditions for vehicle maintenance. Ind. Mark. Manag. **58**(2016), 69–82 (2016)
27. Heskett, J.L., Sasser, W.E., Schlesinger, L.A.: The Service Profit Chain. Free Press, New York, NY (1997)
28. Hsu, C.: Scaling with digital connection: services innovation. In: 2007 IEEE International Conference on Systems, Man and Cybernetics, vols. 1–8, pp. 4057–4061. IEEE, New York (2007)
29. Kamp, B., Parry, G.: Servitization and advanced business services as levers for competitiveness. Ind. Mark. Manag. **60**, 11–16 (2017). https://doi.org/10.1016/j.indmarman.2016.12.008
30. Kindström, D.: Towards a service-based business model—Key aspects for future competitive advantage. Eur. Manag. J. **28**(6), 479–490 (2010)
31. Kindström, D., Kowalkowski, C.: Service innovation in product-centric firms: a multidimensional business model perspective. J. Bus. Ind. Mark. **29**(2), 96–111 (2014)
32. Kowalkowski, C., Gebauer, H., Kamp, B., Parry, G.: Servitization and deservitization: overview, concepts, and definitions. Ind. Mark. Manag. **60**, 4–10 (2017). https://doi.org/10.1016/j.indmarman.2016.12.007
33. Kowalkowski, C., Windahl, C., Kindström, D., Gebauer, H.: What service transition rethinking established assumptions about manufacturers' service-led growth strategies. Ind. Mark. Manag. **45**, 59–69 (2015)
34. Krafft, M., Kumar, V., Harmeling, C., Singh, S., Zhu, T., Chen, J., et al.: Insight is power: understanding the terms of the consumer-firm data exchange. J. Retail. **97**(1), 133–149 (2021). https://doi.org/10.1016/j.jretai.2020.11.001
35. Lampón, J.F., Muñoz-Dueñas, P.: Are sustainable mobility firms reshaping the traditional relationships in the automotive industry value chain? J. Clean. Prod. **413**, Article 137522 (2023). https://doi.org/10.1016/j.jclepro.2023.137522
36. Leminen, S., Rajahonka, M., Wendelin, R., Westerlund, M., Nyström, A.: Autonomous vehicle solutions and their digital servitization business models. Technol. Forecast. Soc. Change **185**, Article 122070 (2022). https://doi.org/10.1016/j.techfore.2022.122070
37. Li, J.H., Lin, L., Chen, D.P., Ma, L.Y.: An empirical study of servitization paradox in China. J. High Technol. Manag. Res. **26**, 66–76 (2015). https://doi.org/10.1016/j.hitech.2015.04.007
38. Lindsay, A., Downs, D., Lunn, K.: Business processes—attempts to find a definition. Inf. Softw. Technol. **45**(15), 1015–1019 (2003)
39. Llopis-Albert, C., Rubio, F., Valero, F.: Impact of digital transformation on the automotive industry. Technol. Forecast. Soc. Change **162**, Article 120343 (2021). https://doi.org/10.1016/j.techfore.2020.120343
40. MacDonald, E.K., Wilson, H., Martinez, V., Toossi, A.: Assessing value-in-use: a conceptual framework and exploratory study. Ind. Mark. Manage. **40**(2011), 671–682 (2011)
41. Matthyssens, P., Vandenbempt, K.: Moving from basic offerings to value-added solutions: strategies, barriers and alignment. Ind. Mark. Manag. **37**(3), 316–328 (2008). https://doi.org/10.1016/j.indmarman.2007.07.008
42. Meredith, J.R.: Automating the factory: theory versus practice. Int. J. Prod. Res. **25**(10), 1493–1510 (1987). https://doi.org/10.1080/00207548708919928
43. MIT Center for Digital Business & Capgemini Consulting: Digital Transformation: A Road-Map for Billion-Dollar Organizations, p. 68 (2011). http://www.capgemini.com/resources/digital-transformation-a-roadmap-for-billiondollar-organizations
44. Mont, O. (ed.): Product-Service Systems: Panacea or Myth. Lund University (2004)
45. Neely, A.: Exploring the financial consequences of the servitization of manufacturing. Oper. Manag. Res. **1**(2), 103–118 (2008). https://doi.org/10.1007/s12063-009-0015-5
46. Ness, D., Swift, J., Ranasinghe, D.C., Xing, K., Soebarto, V.: Smart steel: new paradigms for the reuse of steel enabled by digital tracking and modelling. J. Clean. Prod. **98**, 292–303 (2015). https://doi.org/10.1016/j.jclepro.2014.08.055

47. Oliva, R., Kallenberg, R.: Managing the transition from products to services. Int. J. Serv. Ind. Manag. **14**(2), 160–172 (2003)
48. Osterwalder, A.: The business model ontology—a proposition in a design science approach. Unpublished Thesis from l'Ecole des Hautes Etudes Commerciales. Université de Lausanne (2004)
49. Ostrom, A.L., Bitner, M.J., Brown, S.W., Burkhard, K.A., Goul, M., Smith-Daniels, V., et al.: Moving forward and making a difference: research priorities for the science of service. J. Serv. Res. **13**(1), 4–36 (2010)
50. Pawar, K.S., Beltagui, A., Riedel, J.C.K.H.: The PSO triangle: designing product, service and organization to create value. Int. J. Oper. Prod. Manag. **29**(5), 468–493 (2009). https://doi.org/10.1108/01443570910953595
51. Perez-Moure, H., Lampon, J.F., Velano-Rodriguez, M.E., Rodríguez-Comesana, L.: Revolutionizing the road: How sustainable, autonomous, and connected vehicles are changing digital mobility business models. Eur. Res. Manag. Bus. Econ. **29**, 100230 (2023). https://doi.org/10.1016/j.iedeen.2023.100230
52. Porter, M.E.: What is strategy? Harv. Bus. Rev. **74**(6), 61–80 (1996)
53. Quinn, J.B.: Intelligent Enterprise. Free Press, New York, NY (1992)
54. Quinn, J.B., Doorley, T.L., Paquette, P.C.: Beyond products: services-based strategy. Harv. Bus. Rev. **68**(2), 58–67 (1990)
55. Robinson, T., Clarke-Hill, C.M., Clarkson, R.: Differentiation through service: a perspective from the commodity chemicals sector. Serv. Ind. J. **22**(3), 149–166 (2002)
56. Rothkrantz, L., Boehle, J., van Wezel, M.: A rental system of electrical cars in Amsterdam. Transp. Lett. **5**(1), 38–48 (2013). https://doi.org/10.1179/1942786712Z.0000000005
57. Ruiz de la Torre, A., Sanchez, D.: Evolution of servitization: new business model opportunities. Int. J. Prod. Manag. Eng. **10**(1), 77–90 (2022). https://doi.org/10.4995/ijpme.2022.16719
58. Sawhney, M.S.: Going beyond the product: defining, designing, and delivering customer solutions. In: Lusch, R.F., Vargo, S.L. (eds.) The Service-Dominant Logic of Marketing, Dialog, Debate, and Directions, pp. 365–380. Routledge (2006). 978-0765614919
59. Slywotsky, A., Morrison, D.J.: The Profit Zone: How Strategic Business Design Will Lead You to Tomorrow's Profits. Three Rivers Press, New York, Chichester (2007)
60. Stocker, A., Shaheen, S., Meyer, G., Beiker, S.: Shared automated mobility: early exploration and potential impacts. Road Vehicle Automation, vol. 4, pp. 125–139. Springer, Cham (2018)
61. Storbacka, K., Windahl, C., Nenonen, S., Salonen, A.: Solution business models: transformation along four continua. Ind. Mark. Manag. **42**(5), 705–716 (2013). https://doi.org/10.1016/j.indmarman.2013.05.008
62. Teece, D.J., Linden, G.: Business models, value capture, and the digital enterprise. J. Organ. Des. **6**, 1–14 (2017). https://doi.org/10.1186/s41469-017-0018-x
63. Teece, D.J.: Explicating dynamic capabilities: the nature and microfoundations of (sustainable) enterprise performance. Strateg. Manag. J. **28**(13), 1319–1350 (2007)
64. Terama, J.E., et al.: The contribution of car sharing to the sustainable mobility transition. Transfers **8**(2), 113–121 (2018)
65. Tukker, A.: Eight types of product–service system: eight ways to sustainability? Experiences from SusProNet. Bus. Strateg. Environ. **13**(4), 246–260 (2004). https://doi.org/10.1002/bse.414
66. Valtakoski, A.: Explaining servitization failure and deservitization: a knowledge based perspective. Ind. Mark. Manag. **60**, 138–150 (2017). (in this issue)
67. Vandermerwe, S., Rada, J.: Servitization of business: adding value by adding services. Eur. Manag. J. **6**(4), 314–324 (1988)
68. Vargo, S.L., Lusch, R.F.: Evolving to a new dominant logic for marketing. J. Mark. **68**, 1–17 (2004)
69. Venkatesh, R., Mathew, L., Singhal, T.K.: Imperatives of business models and digital transformation for digital services providers. Int. J. Bus. Data Commun. Netw. **15**(1), 105–124 (2019). https://doi.org/10.4018/ijbdcn.2019010107
70. Visnjic Kastalli, I., Van Looy, B.: Servitization: disentangling the impact of service business model innovation on manufacturing firm performance. J. Oper. Manag. **31**, 169–180 (2013)

71. Wise, R., Baumgartner, P.: Go downstream: the new imperative in manufacturing. Harv. Bus. Rev. **77**(5), 133–141 (1999)
72. Xerox Consulting. https://www.xerox.com/en-us/about/equipment-training-and-consulting. Accessed 25 Feb 2025
73. You, W., Sarpong, D., O'Regan, N.: Global value chains in a digitalised era. Strateg. Chang. **31**(1), 5–8 (2022). https://doi.org/10.1002/jsc.2476
74. Lerch, C., Jäger, A., Meyer, N.: I4.0-Readiness – Baden-Württemberg auf dem Weg zur Industrie 4.0? Studie im Auftrag des Ministeriums für Wirtschaft, Arbeit und Woh-nungsbau Baden-Württemberg. Karlsruhe: Fraunhofer ISI. (2016)https://www.i40-bw.de/wp-content/uploads/2020/08/Studie_I4.0_Readiness_Fh_ISI_publiziert-2_2016.pdf. Accessed 25 June 2025
75. Chesbrough, H.: Open Services Innovation: Rethinking Your Business to Grow and Compete in a New Era, Henry William Chesbrough, John Wiley & Sons, ISBN 0470905743, 9780470905746, pp. 256 (2011)

Chapter 2
The Concept and Operation of Autonomous Vehicles

2.1 Introduction

The subject of this chapter is to explain the concept of 'autonomous vehicle'. This is important for the following reasons: First, autonomous vehicles constitute a key context for servitisation, i.e., transforming business models from those focused on products to those centred on services. Second, autonomous vehicles represent a critical area within the field of artificial intelligence. Therefore, in this chapter, as a starting point for the analysis undertaken, the concept of 'artificial intelligence' is discussed, referring to the legal definitions and viewpoints presented in the literature on the subject. This chapter also encompasses classifications of artificial intelligence and the concept of an 'artificial agent'. The features of an 'artificial agent' distinguished in this chapter facilitate the exploration of connections with the characteristics of an 'autonomous vehicle'.

The latter concept is addressed by exploring legal definitions and the perspectives articulated in the literature. Moreover, the classification of 'autonomous vehicles' is investigated. This is crucial for ascertaining the responsibilities of both the actual driver and the artificial agent in controlling an autonomous vehicle within a defined driving mode. To comprehensively comprehend the function of the artificial agent as a driver of an autonomous vehicle, this chapter examines the fundamental principles of operation. In addition, it considers challenges associated with programming artificial intelligence algorithms intended for steering an autonomous vehicle and establishing priorities concerning the protected goods in relation to the vehicle's movement.

A. Krasuski, *Legal Challenges of a Transformation from a Product-Centric to a Service-Centric Business Model and Procedures*, Studies in Systems, Decision and Control 610, https://doi.org/10.1007/978-3-031-98901-8_2

2.2 Definition of Artificial Intelligence

The determination of the meaning of the term 'autonomous vehicle' necessitates an elucidation of the terms 'artificial intelligence' (AI) and 'artificial agent'. Artificial intelligence represents a collective terminology for various algorithm-based technologies that possess different characteristics and application domains. It has been articulated in terms of its perceived intelligence [65], its capacity to act autonomously [87], and its characteristic of exhibiting evolving behaviour in an unpredictable manner [24]. Consequently, it encompasses a range of algorithm-based technologies with distinct characteristics and designations.

The Report of the European Union High-Level Expert Group on Artificial Intelligence identifies complexity, opacity, openness, autonomy, predictability, data-driven nature, and vulnerability as distinctive characteristics of artificial intelligence [20]. In establishing a suitable liability framework for AI-based technologies it is crucial to assess how these characteristics influence the risks associated with artificial intelligence.

Hallevy [31] identified five attributes that an intelligent entity ought to possess. These are as follows:

(1) Communication;
(2) Internal knowledge;
(3) External knowledge—the entity's awareness of itself and its surrounding environment;
(4) Capability to act in pursuit of a specific goal; and
(5) Creativity.

In the realm of scientific research related to artificial intelligence, expertise from a diverse range of disciplines is essential, including mathematics, logic, statistics, and computer science. More specifically, it refers to specific processes or capabilities carried out by computational devices (generally referred to as personal computers, tablets, robots, etc.) or their respective software [96].

The circumstances and procedures for obtaining the requisite data and tools for the development, testing, and practical application of artificial intelligence encompass a broad spectrum of social, economic, political, and various other relations. The application of AI within these relationships necessitates interaction with the governing legal framework. For artificial intelligence to be subject to legal regulation and accountability regarding its usage, it is essential to clearly define the nature of artificial intelligence, the mechanisms by which it operates, and the potential consequences of its use, both during the testing phase and in general circulation.

The legal regulation concerning the definition of 'artificial intelligence' presents several challenges for legislators. It is not merely a matter of ascertaining the meaning of the concept of 'artificiality'; rather, it involves delineating what is meant by the concept of 'intelligence' [87]. Specifically, this pertains to determining when a particular creation—and, considering subjectivity, its essence—can be regarded as intelligent [62].

The concept of 'intelligent' or 'endowed with intelligence' is inherently human. Thus, defining intelligence in this manner relies significantly on the perception and evaluation of the fulfilment of various 'human' attributes and characteristics. In assessing the manifestation of intelligence, comparisons may consequently be drawn to human behaviour; specifically, an artificial agent endeavours to (1) think like a human, (2) act like a human, (3) ratiocinate, and (4) behave rationally [87].

In succinct terms, the first two categories of anthropocentric approaches are based on and utilise the findings of empirical sciences, primarily the human sciences and experimental results, while the third and fourth categories highlight rational behaviour and are rooted in the methodologies and practices of mathematics, logic, and technical sciences [80].

Learning acts as a prerequisite for the development of human intelligence, a principle that similarly applies to artificial intelligence. Artificial intelligence primarily employs two methods of 'learning': machine learning and knowledge representation and reasoning. Machine learning is a subfield of artificial intelligence that focuses on methodologies and algorithms enabling a programme (artificial intelligence) to learn and subsequently respond appropriately to varying input values without being explicitly programmed. It functions solely based on the information it has assimilated. Machine learning algorithms integrate elements of mathematical statistics, statistical analysis techniques, and data analysis. They can address three fundamental problems (tasks): classification, regression, and clustering. Algorithms are categorised into three types based on the learning approach: supervised learning, unsupervised learning, and reinforcement learning [89].

Knowledge representation and reasoning constitute a vital domain within the field of artificial intelligence, encompassing the methodologies through which knowledge is symbolically articulated and automatically processed by knowledge-based systems (programs) [11]. The concepts of knowledge representation and reasoning, along with their respective methodologies, predominantly rely on logical frameworks and techniques aimed at formally encoding knowledge into a format suitable for computer processing. Furthermore, knowledge representation and reasoning draw upon psychological insights concerning how individuals approach problem-solving and knowledge representation, thereby proposing formalisations that facilitate the design and development of intricate systems. A quintessential example of artificial intelligence employing knowledge representation and reasoning is expert systems, or, more broadly, knowledge-based systems. Alternative methodologies for achieving 'learning' in artificial intelligence encompass statistical methods, Bayesian estimation, among others.

It is also important to highlight the categorisation of artificial intelligence into 'weak' (narrow) and 'strong' (general) artificial intelligence. This classification is based on the criterion of actions that are independent of human intervention. Weak artificial intelligence possesses 'intelligence' confined to a specific domain, within which the system is capable of repeatedly solving programmed tasks or a singular task under constant human oversight [49]. This system can perform independent assessments and make decisions within the confines of its design and programming; however, it lacks the capacity for autonomous reasoning and control. It embodies the intentions of the creator or user of the system. Strong AI is characterised by its capacity to identify and exert control. It possesses the ability to make independent decisions and choices within the parameters of its design and programming, effectively implementing the will of the creator or user. Moreover, it transcends its designed framework, enabling it to make autonomous decisions and execute appropriate behaviours in alignment with its own objectives [101]. In this manner, strong AI addresses challenges independently without the necessity for human intervention.

In academic discourse, a third category of artificial intelligence is identified, namely, artificial super-intelligence, which would surpass human intelligence and capabilities to such an extent that even the most exceptional human intellects would be unable to compete with it.

Müller [67] utilises the term 'singularity'. Beyond this juncture, artificial super-intelligence systems possess the capacity to enhance and generate even more refined systems, the behaviours of which may prove challenging for humans to anticipate.

The classification of artificial intelligence into categories of weak and strong is illustrated by the levels of automation established by the Society of Automotive Engineers [83]. According to the SAE [84], vehicle automation is categorised into six levels, ranging from 0 to 5. Vehicles classified at levels 1 and 2 utilise what are referred to as weak AI systems, whereas vehicles classified at level 3 and above operate or embody a strong AI system that processes information (input) and makes autonomous decisions based on this information as well as on the outcomes of 'learning' in traffic conditions [37].

Artificial intelligence can be defined by examining the relationship between the level of detail of the defining phrases (lat. definiens) used to describe the defined phrase (lat. definiendum) and the time frame during which the definition remains valid. Based on this criterion, it becomes possible to differentiate between ex ante and ex post definitions of artificial intelligence.

In the ex-ante approach to defining artificial intelligence (AI), the objective of the definer, such as the legislator, is to establish a definition that considers all potential techniques—both those currently utilised and those anticipated for future use. This definition aims to be resilient against the evolution of new technologies over the long term. The existing literature [96] indicates that such an approach poses the risk of the definition being inadequate, overly general, or ambiguous. A significant challenge arises when the progression and characterisation of AI diverge from the predictions made by the legislator, potentially hindering their ability to respond adequately to these emerging changes with sufficient flexibility.

An illustration of an ex-ante definition is the definition of 'artificial intelligence system' as articulated in Article 3(1) of the proposal for a Regulation of the European Parliament and of the Council, which establishes harmonised rules concerning artificial intelligence (AI Act) and amends certain legislative acts of the Union (COM(2021) 206 final) [13]. The term 'artificial intelligence system' is delineated as software developed utilising one or more of the techniques and approaches enumerated in Annex I, which can, for a specified set of human-defined objectives, produce outputs such as content, predictions, recommendations, or decisions that influence the environments with which it interacts. Annex I catalogues three specific techniques that pertain to the notion of 'artificial intelligence':

(1) 'machine learning approaches', including supervised, unsupervised and reinforcement learning, using a wide range of methods, including deep learning;
(2) 'logic and knowledge-based approaches', including knowledge representation, inductive (logical) programming, knowledge bases, inference and deductive engines, reasoning (symbolic), and expert systems; and
(3) 'statistical approaches' including Bayesian estimation, search and optimisation methods.

Recital 6 of the Proposal for a Regulation on Artificial Intelligence indicates, among other matters, that the concept of an artificial intelligence system should be distinctly defined to ensure legal certainty while allowing for the flexibility to adapt to future technological developments. Additionally, an example of an ex-ante definition is provided in the definition of artificial intelligence, as addressed in Article 3 (1) of Regulation (EU) 2024/1689 (AI Act, 2024) [77]. The AI Act aims to 'improve the functioning of the internal market and promote the implementation of human-centric and trustworthy artificial intelligence (AI)' [19]. Certain positions in the literature have suggested that the original definition of artificial intelligence is overly broad [69], encompassing a wide range of algorithmic technologies, from Bayesian estimation to deep learning [41].

The AI Act delineates artificial intelligence as a machine system engineered to function with varying degrees of autonomy subsequent to its deployment. It possesses the capability to adapt post-deployment and, for explicit or implied purposes, derives outputs from received inputs—such as predictions, content, recommendations, or decisions—that may influence a physical or virtual environment.

An ex-post definition accurately reflects the current state of knowledge regarding new technologies at the time it is formulated. Conversely, this type of definition is vulnerable to changes within the technology field, which presents the risk of not adapting to technological advancements.

When a legislator modifies the content of a definition in response to contemporary circumstances regarding artificial intelligence, he or she consequently alters the approach to defining artificial intelligence from ex-ante to ex-post [33].

The definition of artificial intelligence may also be contingent upon the objectives of the relevant regulation. In this context, it is possible to differentiate between definitions established for general and sectoral applications. Universal (systemic, general) definitions articulate the concept of artificial intelligence, capturing its essence while excluding the outcomes of its application. According to such a definition, one can determine whether a particular system (or product) qualifies as part of artificial intelligence systems and the overarching regulations pertaining to artificial intelligence. A pertinent example is provided by the Act on AI, which delineates a definition of an 'AI system'.

Product-sector (ad hoc) approaches are characteristic of specific, fragmented regulations within particular sectors. This approach does not precisely define artificial intelligence. The subject of the definition pertains to an entity (e.g., an object) that utilises or implements artificial intelligence, fundamentally influencing its activities and operations. Consequently, the results predominantly consist of definitions that characterise the product as fully or partially autonomous, or as capable of performing tasks independently. In this manner, these definitions essentially reflect the outcome of the application of artificial intelligence (autonomous behaviour) [86].

2.3 Definition of an Artificial Agent

The notion of an artificial agent currently lacks a legal definition. In the context of this monograph, it is used to illustrate the intersection between the idea of an autonomous vehicle and the concept of artificial intelligence.

The concept of an artificial agent encompasses the following elements.

Firstly, it is an information technology system (IT); specifically, a collection of software. It operates within a designated operating system, incorporating middleware and other applications. This system enables the artificial agent to behave intelligently (i.e., to act or refrain from acting) as articulated by Kirn et al. [49], allowing the agent to autonomously engage in a learning process aimed at modifying its previous behaviour. Such modifications are evaluated within the context of the environment in which the artificial agent operates. Furthermore, this capability allows the artificial agent to independently utilise resources available on the Internet, such as computational capacity, computing power, and data transfer capabilities between servers. These functionalities are facilitated through services in cloud computing [53] and virtualisation [55]. The system is unlimited in temporal scope, yet its reach is confined to the extent of Internet connectivity and access to online services.

The concept of 'autonomous software modules' (German: autonome Softwaremodule), introduced by Muller [66], does not alter the meaning of the above description.

Secondly, the information technology system operates according to established action objectives [18] and is equipped with 'rewards', which are defined as the successful achievement of these objectives.

Thirdly, the primary aim of utilising the functionalities provided by an IT system is to strive for the ability to replace human involvement in executing tasks. More specifically, artificial agents may be employed to replicate animal behaviours. In this regard, it is essential to assess the feasibility of mimicking human behaviours and, consequently, substituting them, particularly when the IT system possesses the capacity to emulate human emotions, including profound sentiments such as pain and joy. Kennedy [45] described this attribute as human grounding.

Fourthly, the artificial agent uses one or more machine learning techniques to carry out specific tasks [51].

Fifthly, the information technology system, described as an intelligent system, excels in achieving objectives pertinent to various environments [57]. Rational actions are taken concerning the methodologies used to realise these objectives [10].

Sixth, a self-learning information technology system can function independently, known as a "software agent." It can be stored and operated on different devices, such as smartphones [60].

It can also serve as a control element of a device, collectively referred to as a "robot". Consequently, the concept of an artificial agent functions as a collective term for both a software agent and a robot [30]. This method of differentiation is also prevalent in the scientific discourse surrounding this field [63].

2.4 Definition of an Autonomous Vehicle

In the literature pertaining to the subject, as well as in the media, a variety of terms are employed to characterise a motor vehicle capable of operating without the involvement of the driver. These terms include: 'autonomous vehicle' [102], 'self-propelled vehicle', 'driverless vehicle', 'computer-controlled cars', and 'wheeled mobile robots'. The monograph utilises a singular term: autonomous vehicle. To ascertain its meaning, it is essential to consider the subsequent issues, presented in the form of processes, including:

(1) The procedure for determining the mode of transport for travel by vehicle is delineated.
(2) The procedure for organising a journey via vehicle encompasses the specification of the departure date and time, the anticipated arrival, and the selection of the travel route, inclusive of any planned stops.
(3) The execution of the journey according to the plan established in point 2 is subsequently addressed.

The processes referenced above pertain to specific behaviours, and their analysis necessitates consideration of both objective and subjective factors.

The objective aspect arises from the content inherent in each of the aforementioned activities. Conversely, the subjective aspect necessitates an analysis of the individuals executing these activities and those overseeing their execution.

In the context of the subjective aspect of the undertaken processes, the etymological meaning of the concept of the automobile can be understood more broadly as a 'self-driving car' [58]. A self-driving car can be interpreted in two distinct ways: firstly, as a mode of transportation that permits an individual to travel without reliance on the muscle power of horses (the originally accepted understanding); secondly, as an artificial agent, specifically a self-learning machine. From this perspective, the term "self-driving" in relation to the vehicle is conceptualised independently of the human who previously executed (before the self-learning machine was operational) the necessary functions to operate this vehicle.

In conclusion, the concept of the 'automobile' can also be applied to vehicles that operate without human involvement in the process of controlling them and making decisions in this area [71].

In ascertaining the meaning of the concept of an 'automobile', it is essential to delineate the subjective aspect associated with the processes outlined at the outset. This necessitates identifying the delineation of activity between a human and an artificial agent, such as a self-learning machine.

The concept of a self-learning machine, within the context under consideration, should be comprehended as software that:

(1) has been installed directly within a specified (dedicated) vehicle or at a distinct location that facilitates the execution of the tasks outlined in items 2–6 below;
(2) governs the assistance systems integrated into the vehicle;
(3) manages the initiation and cessation of the vehicle's operation;
(4) oversees the vehicle's functions concerning: (a) executing vehicle manoeuvres, including moving straight ahead and altering the direction of travel, (b) adhering to road traffic regulations, (c) preventing collisions with other vehicles and pedestrians;
(5) acquires knowledge on road behaviour in accordance with a specific conduct pattern by gathering data from sensors installed within the vehicle;
(6) allows for modifications in the actions performed as indicated in items 2–4, as a result of the learning effect referred to in item 4.

A self-learning machine assumes the role of a human driver, taking over the driving functions of the vehicle for the entire duration of the journey or a specified segment thereof. The second and third processes mentioned at the outset of this subsection may be executed by an autonomous vehicle, wherein the role of the driver would be fulfilled by an artificial agent rather than a human. The inference regarding the capability of an artificial agent, such as a self-learning machine, to carry out the second and third processes arises from its endowment with functionalities that align with the definition of a self-learning machine. It can be contended that reallocating the capacity for vehicular travel from a human to an artificial agent may constitute an infringement of individual civil rights. For instance, in Poland, civil rights have been enshrined in the Polish Constitution (1997) [93], which guarantees freedom of movement as stipulated in Article 56, paragraphs 1–3. Conversely, one could conceive of transferring the authority to make decisions regarding vehicular travel from a human to an artificial entity. This transition may occur when, based on the administrative decision rendered by the district governor to revoke the right to operate vehicles, among other circumstances such as the determination of an individual's incapacity to drive, infractions committed by the driver, and judicial rulings, a specific individual is prohibited from operating motor vehicles. Such a circumstance stems from Article 103(1) of the Polish Act on Vehicle Drivers (2011) [92]. The actions of an artificial agent that decide to deny an individual the ability to undertake the intention of travelling by motor vehicle must be regarded as lawful. This would fall under the category of overseeing the enforcement of a penal sanction imposed on the accused, specifically a prohibition against driving any motor vehicles in land traffic. Moreover, the actions of an artificial agent that decide to deny an individual the ability to undertake the intention of travelling by motor vehicle must be regarded as lawful. This would fall under the category of overseeing the enforcement of a penal sanction imposed on the accused, specifically a prohibition against driving any motor vehicles in land traffic (Article 43 § 1 of the Polish Penal Code) [2].

The literature presents definitions of equality that are based on a singular characteristic of an autonomous vehicle, namely its ability to operate independently of human intervention, particularly in steering. Predominantly, these definitions characterise an autonomous vehicle from a functional perspective [58]. Conversely, other definitions delineate an autonomous vehicle from a subjective standpoint, defining it as a robotic vehicle that functions without a human operator [8, 44, 72].

This operation concerns the execution of specific control functions that are crucial for ensuring safety regarding the movement of the vehicle, which include: (1) determining the direction of travel; (2) determining the speed of travel through the regulation of the accelerator pedal and control of engine speed; and (3) braking through the manipulation of the brake pedal. These functions are carried out without direct participation from the driver [68].

The perception of an autonomous vehicle discussed in this monograph is supported with specific reservations. The vehicle's autonomy, defined as the capacity to operate without human intervention, arises from the implementation of a control module for support systems, known as an automated driving system (ADS). This control module aligns with the characteristics of an artificial agent. The ADS holds internal significance, pertaining to the vehicle it governs, as it facilitates the operation of this vehicle independently of human involvement. Additionally, it possesses external significance, as it enables the integration of individual autonomous vehicles with one another into a 'column of vehicles,' often referred to as a 'platoon.' The process of connecting vehicles is termed 'platooning.' A consequence of establishing a 'platoon' is the maintenance of a constant speed in alignment with the leading vehicle, ensuring a precise distance is upheld between vehicles in accordance with safety regulations. The pivotal aspect of this model of autonomous vehicle movement within a convoy is the autonomy of the leading vehicle, reliant on real-time data processing derived from other connected vehicles [85].

The European Parliamentary Research Service ('EPRS') defines an 'automated vehicle' as a motor vehicle equipped with technology that assists the driver by transferring certain driving functions to a computer system. This indicates that the operator continues to retain a degree of control and responsibility. In contrast, EPRS characterises an 'autonomous vehicle' as a more advanced form of automation, capable of performing all driving tasks independently, that is, without any human intervention [74].

In Regulation 2019/2144, the European Union legislator has delineated the definitions of an 'automated vehicle' (refer to Article 3(21) of Regulation EU 2019/2144) [76] and a 'fully automated vehicle' (refer to Article 3(22) of Regulation EU 2019/2144) [76]. The former definition pertains to a motor vehicle that is engineered and fabricated to operate autonomously for a specified duration without continuous oversight by a driver, however, driver intervention is still anticipated or required. Conversely, the latter definition pertains to a motor vehicle that is engineered and fabricated to operate autonomously without any oversight from a driver.

The draft amendment to the proposal concerning the 1968 Convention on Road Traffic [79], prepared by the Global Road Safety Forum of the United Nations Economic Commission for Europe in 2020 (hereinafter referred to as 'CRD'), introduced two new definitions into Article 1: (ab) 'Automated driving system' denotes a vehicle system that utilises both hardware and software to continuously exercise dynamic control of a vehicle; (ac) 'Dynamic control' pertains to the real-time execution of all operational and tactical functions necessary for vehicle movement. This encompasses the control of both the lateral and longitudinal motion of the vehicle, monitoring road conditions, responding to traffic incidents, and planning and signalling manoeuvres. Furthermore, the proposed amendment incorporated Article 34 CRD, which stipulates: 'The requirement for each moving vehicle or combination of vehicles to have a driver is considered satisfied when the vehicle employs an automated steering system that adheres to: (a) national technical regulations and any pertinent international legal instruments related to wheeled vehicles, their equipment, and parts that may be installed and/or utilised on wheeled vehicles, and (b) national regulations governing operational practices'. The amendments to Article 1, along with the new Article 34 bis of the CRD, were accepted by January 2022 and subsequently came into effect on 14 July 2022.

An additional example is the UK Automated and Electric Vehicles Act [6], which represents one of the pioneering European regulations that refers to automated vehicles as those 'designed or adapted to be capable, at least in some circumstances or situations, of safely driving themselves and to be lawfully used while driving themselves, at least in some circumstances or situations, on a road or other public place in the United Kingdom.' Furthermore, it specifies that 'a vehicle is driving itself if it is operating in a mode in which it is not controlled by, and does not need to be monitored by, a person.'

In Article 65k of the Polish Road Traffic Act, the Polish legislator defines an 'autonomous vehicle' as a motor vehicle that is equipped with systems capable of controlling its movement, thereby enabling the vehicle to operate independently of driver intervention, although the driver retains the capacity to regain control at any moment. Within this definition, the legislator identifies two sources of control for an autonomous vehicle: 'systems controlling the movement of this vehicle' and 'the driver', as defined in Article 2(20) of the Polish Road Traffic Act. This definition reveals a logical inconsistency; the vehicle is no longer classified as autonomous once control is relinquished from the 'systems controlling the autonomous vehicle'. Consequently, the definiens fails to adequately define the definiendum of this concept, as established by the legislator's definition of equality. Furthermore, the terminology 'control systems' is misleading. Different systems cannot simultaneously manage the vehicle's operations; conflicting commands would inevitably arise due to the existence of multiple 'controls'. It is imperative to designate a singular system responsible for the vehicle's motion.

An autonomous vehicle consists of a suite of devices, including sensors and software, specifically designed to perform tasks associated with various levels of automation. These tasks include: (1) partial automation, (2) conditional automation, (3) high automation, and (4) full automation. The array of devices and software is connected to the Automated Driving System (ADS). At automation levels ranging from 3 to 5, the ADS functions as an artificial agent, akin to a driver (robot) executing tasks based on principles of machine learning. In the academic literature, the term 'machine driver' is utilised to denote this artificial agent.

The terms 'autonomous' and 'automated' in relation to a vehicle should not be construed as synonyms. 'Autonomous' signifies self-governing, whereas 'automated' denotes automatic functionality.

The concept of autonomy can be analysed through the following characteristics (Sartor, 2016):

(1) Independence—referring to the system's capacity to independently execute tasks (agential independence) and organisational freedom, which denotes the system's engagement in tasks within the broader socio-technical framework.
(2) Cognitive skills—denote the system's capability to carry out complex discriminative functions. This encompasses information acquisition, analysis, decision-making, implementation, and cognitive architecture, which includes the system's goal orientation, adaptability, and capacity to represent the environment, establish objectives to accomplish, and identify suitable means to achieve those objectives.

    A distinction between 'autonomous driving' and 'automated driving' can also be made based on the definitions provided in the EU SMART Report64 [17]:
(3) Automated driving refers to driving that is facilitated by specialised controls, as well as existing autonomous subsystems that assist the driver in taking or regaining control of the vehicle promptly. The driver remains legally responsible for executing the driving task.
(4) Autonomous driving is defined as the outcome of automated driving. In principle, there is no requirement for the driver to be actively engaged in operating the vehicle, although the driver may or may not still occupy the driver's seat [70].

For further considerations, the concept of an autonomous vehicle is predicated upon the following features:

(1) a vehicle that functions as a mechanical means of transportation powered by natural forces,
(2) equipped with Automated Driving Systems (ADS), which enable the integration and management of support systems installed within the vehicle,
(3) resulting in the capability to perform the task of steering the vehicle autonomously, without the involvement of a human operator, whether inside or outside the vehicle (internal meaning), as well as the ability to interact with other autonomous vehicles (for example, by forming a convoy of such vehicles) (external meaning) [54].

The term 'autonomy' pertaining to an information system should be interpreted as the attribute of that system to alter its domain of use or purpose without external intervention, control, or supervision [38].

In the context of the aforementioned definition, it is imperative to emphasise that every autonomous vehicle will function as an automated vehicle; that is, a vehicle equipped with a driving assistance system. Issues concerning privacy protection must be considered from two distinct perspectives. The etymological significance of the term 'autonomy' can be traced back to the ancient Greek concepts of 'autos', meaning 'self', and 'nomos', meaning 'law'. The phrase meaning 'one's law' was interpreted by the ancient Greeks to denote 'making one's law' [27]. Autonomy was associated with the authenticity of an individual as the author of one's actions and with the notion of free will, which is essential to personal responsibility. The Stoics and Epicurus are acknowledged for refining the ancient concept of autonomy into the modern understanding of self-determination, a principle that remains fundamental to personal responsibility. Thomas Aquinas [73] and Immanuel Kant [42] critically examined the concept of 'autonomy' within their frameworks of 'individual agency' and moral responsibility.

In defining an 'autonomous vehicle', the term 'autonomous' refers to the attributes of this vehicle that enable it to navigate without human intervention. Consequently, a human may delegate certain decisions related to the steering of an autonomous vehicle to an Automated Driving System ('ADS'). In this context, an autonomous vehicle, specifically an ADS, operates as an artificial agent [30]. Its responsibility is to make specific choices or decisions that are limited to particular matters. Therefore, the term "artificial agent" serves as a collective descriptor for both 'software agent' and 'robot'. For instance, a vehicle can manage certain functions, such as selecting speed or route, while respecting the decisions made by a human regarding the destination of the journey [21].

Based on the definition provided, it is imperative to emphasise that all autonomous vehicles are classified as automated vehicles, indicating that they are equipped with a driving assistance system. However, it must be noted that not every safety feature qualifies a vehicle as 'automated'. First, systems such as electronic stability control ('ESC') and automatic emergency braking ('AEB'), along with certain driver assistance features like lane keeping assist ('LKA'), fall outside the framework of driving automation taxonomy because they do not continuously engage in the 'Dynamic Driving Task' ('DDT'), but instead offer brief interventions during potentially hazardous situations. Since their actions are momentary, active safety systems do not modify or absolve the driver of the responsibility to execute part or all of the DDT. Consequently, these systems are not classified as driving automation, even if they perform automated functions. Second, systems that alert, warn, or inform the driver of hazards in the driving environment are also excluded from the driving automation taxonomy, as they do not automate the DDT nor alter the driver's role in carrying out the DDT. However, collision avoidance features, including intervention-type active safety systems, may be integrated into vehicles possessing any level of driving automation. For Automated Driving System ('ADS') functions (i.e., levels 3–5) that manage the full DDT, collision mitigation and avoidance functionalities are essential components of the ADS capabilities [75].

2.5 Classification of Automated Vehicles

The Society of Automotive Engineers [83] has presented six levels of motor vehicle automation that continuously perform all DDT. This term should be understood to encompass all real-time operational and tactical functions required to operate a vehicle in traffic, excluding processes involving trip planning and the selection of destinations and waypoints. The automation scale assumes the absence of any automation as a starting point, which is reflected in level 0. The level of automation increases to level 5, corresponding to full automation. These levels are applied to motor vehicles, with each level of automation marked by an appropriate name. Table 2.1 presents the levels of automation for motor vehicles.

Vehicles classified with automation levels ranging from 0 to 2 are equipped with driver support features, while those in automation levels 3–5 are outfitted with driver-automated capabilities and can operate in an autonomous mode. It is important to note that although a specific vehicle may have a driving automation system capable of providing various automation functions at different levels, the precise level of automation depends on its activation.

Level 0 denotes the absence of driving automation, indicating that a human operator is responsible for all driving functions. Certain features, such as automatic emergency braking, blind spot warning, and lane departure warning, are designed to assist the driver. Level 1 encompasses driver assistance, which includes features that aid in steering, braking, and acceleration, for example, through lane centring or adaptive cruise control [75].

Level 2 pertains to partially automated driving. The vehicle is equipped with driver-assistance features, including steering, braking, and acceleration assistance, which encompass the simultaneous execution of lane centring and adaptive cruise control. The system installed within the vehicle is capable of managing both lateral and longitudinal movement. However, the driver is required to carry out the remaining dynamic driving tasks, such as object recognition, and must be prepared to intervene when necessary (SAE, 2018). The driver must remain poised to regain control of the vehicle's longitudinal and lateral movements at all times [83]. The available vehicle models on the market that incorporate automated driving features predominantly fall under Level 2: Tesla's Autopilot, characterised as "full autonomous driving;" Audi's

Table 2.1 The levels of automation of motor vehicles

Levels of automation	Description
Level 0	No driving automation
Level 1	Driver assistance
Level 2	Partial driving automation
Level 3	Conditional driving automation
Level 4	High driving automation
Level 5	High driving automation

Source Original creation

traffic jam assist; GM's Super Cruise; BMW's extended traffic jam assist; Ford's Blue Cruise; and Hyundai's automated driving package [88].

From the perspective of the research issue regarding accountability for actions and omissions associated with the operation of an autonomous vehicle, the transition between level 2 and level 3 automation is significantly important. At level 3, the Automated Driving System (ADS) takes control of tasks related to steering the vehicle, thereby altering the assessment of the driver's accountability. Level 3 encompasses conditional driving automation. In this scenario, the vehicle has features that allow it to operate under specific conditions and does not function autonomously if these prerequisites are not met. The driver must remain attentive and, if necessary, regain control at the request of the ADS. Within the scope of automation level 3, the system can execute the complete Dynamic Driving Task (ddt), yet requires the user to re-establish control of the steering functions when faced with a situation that exceeds the system's capabilities or when the vehicle approaches the limits of its Operational Design Domain (ODD; e.g., the conclusion of a motorway) [83]. The system is expected to issue a request for intervention and provide the user with sufficient time to assume driving responsibility [83].

Certain automotive manufacturers have announced their intention to forgo the production of Level 3 [98] vehicles, citing the risks inherent in transferring the execution of the dynamic driving task from the vehicle's automated driving system to the user. Conversely, other manufacturers are advocating for the widespread introduction of Level 3 vehicles [5]. A notable example is Mercedes-Benz, whose 'Drive Pilot' was recognised as the world's first fully certified Level 3 autonomous driving system in 2022 [92]. This system facilitates driving at speeds of up to 60 km/h within designated zones. The driver may disengage from the steering wheel while the automated driving system assumes control of the vehicle's functions. The manufacturer of Mercedes-Benz vehicles has asserted that it will assume legal liability for accidents directly attributed to errors made by the automated driving system. In instances where an accident results from the driver's negligence, liability for damages will rest with the driver [52].

In contrast to Level 3, a Level 4 vehicle system is capable of executing the complete dynamic driving task within its Operational Design Domain (ODD) without requiring human intervention, such as on a motorway [83]. A Level 5 vehicle represents the pinnacle of automation, as the Automated Driving System (ADS) can manage the entire dynamic driving task in any environment, across all types of roads and under varying weather conditions. To facilitate autonomous driving and ensure safety functions, the vehicle collects substantial amounts of data. This level of automation most accurately encapsulates the idea of a self-driving (autonomous) vehicle. In certain instances, there may not even be an occupant present within the vehicle. Accordingly, this level of automation, designated as Level 5, is commonly referred to as "full automation" [81].

The aforementioned issues should be assessed within a subjective framework. This relates to the connection between a human (driver) and the autonomous driving system (ADS), as well as support systems and other vehicle components, all of which impact the act of driving, including what is referred to as 'autonomous' driving.

2.6 Operational Modes of Autonomous Vehicles

Autonomous vehicles are equipped with an array of sensors designed to collect data about the surrounding environment in which the vehicle operates (in motion) (Gates, 2016). Cameras and radar systems are often used to detect nearby objects and determine the vehicle's distance from them. Additionally, Light Detection and Ranging units (LIDAR) may be employed; these devices generate a scanned 360° image of the vehicle's environment using laser rangefinders [93].

Software algorithms then interpret this data and control the vehicle's course.

Autonomous vehicles are equipped with a range of sensors that collect data about the environment in which the vehicle operates while in motion (Gates, 2016). Cameras and radar systems are often used to detect nearby objects and determine the vehicle's distance from them. Additionally, Light Detection and Ranging units (LIDAR) can produce a comprehensive 360° scan of the vehicle's surroundings using laser rangefinders [93]. Thereafter, software algorithms analyse this data to regulate the vehicle's trajectory.

The process through which algorithms guide a vehicle can be succinctly categorised into three stages: (1) vehicle localisation; (2) obstacle detection and behaviour prediction; and (3) path planning [39].

The automated driving system initiates its operations by ascertaining its geographic location through the utilisation of GPS and an inertial navigation system (INS), which employs gyroscopes and accelerometers to evaluate the vehicle's position, orientation, and speed [4].

Localisation and mapping algorithms determine the global and local position of the vehicle while mapping the surroundings using sensor data and other perception outputs [64].

Path planning identifies feasible and safe routes for a vehicle, utilising perceptual data, positional information, and mapping analytics [43]. The decision-making component computes the optimal route by evaluating potential paths, the vehicle's current status, and relevant environmental data (e.g., road characteristics, meteorological conditions, traffic signage, etc.). Subsequently, the vehicle control module formulates the necessary commands (torque, acceleration, steering angle, etc.) to implement the optimal route decision, which may include lane changes, right turns, or other driving manoeuvres [29]. It is essential to note that the process of autonomous navigation functions as a high-frequency recursive procedure [9]. This characteristic empowers autonomous vehicles to navigate competently through high-speed traffic and interact with dynamic entities such as pedestrians, motorcycles, and automobiles [40].

The driver sets the destination, prompting the software to calculate the route and initiate travel. A rotating LIDAR sensor mounted on the vehicle's roof monitors a 60-m radius around the automobile and generates a dynamic three-dimensional map of the vehicle's immediate environment. A sensor located on one of the wheels assesses lateral movement to determine the vehicle's position in relation to the three-dimensional map. Radar systems situated in the front and rear bumpers measure the distance to nearby obstacles. The artificial intelligence (AI) software within the vehicle is connected to all sensors and utilises data from Google Street View and video cameras. The AI mimics human perception and decision-making processes, managing critical driving systems such as steering and braking. The automated driving system (ADS) interfaces with Google Maps to provide advanced notifications regarding landmarks, road signs, and traffic lights. An override mechanism is available, allowing a human operator to regain control of the vehicle. Individual vehicles can utilise data from nearby approaching vehicles, especially information regarding traffic congestion and potential safety hazards. In-vehicle communication systems utilise nearby vehicles and roadside devices as communication nodes within a peer-to-peer network (Shen, 2010), enabling the exchange of information [71].

Autonomous vehicles utilise a variety of techniques to perceive their environment, including radar, lasers, GPS, odometers, and computer vision [35]. Advanced control systems are responsible for interpreting information gathered from installed sensors to identify suitable navigation routes, as well as obstacles and relevant signage. These vehicles are equipped with control systems capable of analysing data from such sensors to differentiate between various vehicles on the road, which proves advantageous in plotting a course to a specified destination [4].

Numerous additional techniques can be employed to ascertain the vehicle's position. For instance, both a camera [26] and LIDAR [34] can detect lane markings and accurately position the vehicle. Moreover, companies such as Ford [97] and Google [39] utilise methods that compare sensor data with "pre-made maps" to enhance the precision of the vehicle's location on the roadway [39]. In the subsequent stage, the vehicle's sensors identify potential obstacles along its route, including other vehicles, cyclists, pedestrians, and road construction sites [39].

The algorithms utilised by the driving system, which are fundamentally based on advanced machine learning programming methodologies and probabilistic reasoning [93], categorise these objects and anticipate the potential behaviour of obstacles. For instance, the Autonomous Driving System (ADS) integrated within Google's automated vehicle prototypes has been "trained" to identify cyclists utilising camera footage and to discern that a cyclist whose arm is extended to the side is likely indicating an intention to turn [39].

However, since manufacturers of automated vehicles are unable to develop algorithms that can recognise all conceivable events occurring within the driving environment, it is essential that these vehicles are also programmed to accommodate atypical driving situations. Below is a logical diagram illustrating the planning and execution of a trip using an autonomous vehicle.

In exceptional circumstances during road traffic, where the Automated Driving System (ADS) cannot identify or anticipate the behaviour of an object, the autonomous vehicle will either proceed with caution or come to a complete stop, depending on the clarity of the situation. At this point, machine learning methodologies are reapplied to inform the vehicle's actions based on overarching priorities.

For instance, the prototype of Audi's automated vehicle bases its decision-making on two fundamental questions: (1) Is it feasible (safe and legal)? and (2) Is it advantageous (for a more comfortable driving experience)? This methodology enables the vehicle to progress while avoiding obstacles and complying with traffic regulations, such as responding to alterations in traffic signals detected by the camera sensor.

The operation of autonomous vehicles encompasses their ability to stop in order to attain a state of minimal risk. According to the Society of Automotive Engineers (SAE), this state involves the vehicle coming to a stop, which can be facilitated by a human driver or an Automated Driving System (ADS) following the performance of Dynamic Driving Tasks (DDT) during emergency situations, thereby mitigating the risk of an accident when a particular journey cannot or should not be continued [83]. Fundamentally, the minimum risk state represents a secure condition that a vehicle can achieve when it is neither feasible nor safe to proceed with its journey.

In the context of automation levels one and two, it is posited that the driver (human) will attain a state of minimal risk when deemed necessary. Regarding automation level three, it is presumed that the user, prepared to execute emergency Dynamic Driving Tasks (DDT), will achieve a state of minimal risk when it is essential or will carry out DDT through alternative methods if the vehicle is operational [16].

Within the parameters of automation levels 4 and 5, the Automated Driving System (ADS) can autonomously achieve a state of minimal risk when necessary, for example, in the case of an Operational Design Domain (ODD) exit or a system failure that affects the Dynamic Driving Task (DDT). Table 2.2 outlines the technology used in autonomous vehicles.

According to SAE [84], the concept of the automated driving system (ADS) should be understood as encompassing both hardware and software capable of executing the entire dynamic driving task continuously. Furthermore, this concept specifically relates to the following levels of driving automation: 3, 4, and 5.

From a technical perspective, the Automated Driving System (ADS) requires various technical functions for its operation, including sensors, radar systems, global positioning systems, and onboard cameras for navigation and environmental sensing, as well as servos for controlling brakes, steering, acceleration, and the gearbox. The data collected by the ADS is recorded by advanced control systems [47], which make informed decisions regarding navigation and detected obstacles. The system must interpret traffic signals and safely manoeuvre among other vehicles to guide the vehicle to its intended destination [36].

In the context of autonomous vehicles, Automated Driving Systems (ADS) can execute all facets of Dynamic Driving Tasks (DDT) continuously without requiring driver intervention under specific circumstances (Levels 3 and 4) and under all circumstances (Level 5) [102]. The primary distinction between a Level 3 and a Level 4 autonomous vehicle lies in the requirement that the driver of a Level 3 autonomous vehicle must be prepared to regain control whenever prompted by the ADS, whereas the driver of a Level 4 autonomous vehicle is not subject to this requirement, as the ADS can achieve a state of minimal risk, thereby allowing the vehicle to operate autonomously.

Table 2.2 Technologies employed in autonomous vehicles

Name	Abbreviation	Description
Light detection and ranging	LIDAR	This technology pertains to remote sensing and facilitates distance measurement by illuminating a target with a beam of light and subsequently analysing the reflected light. A compatible pod is affixed to the vehicle's roof, situated within a cylindrical housing that rotates a full 360°. LIDAR is comprised of a transmitter, a mirror, and a receiver, enabling the mapping of the three-dimensional structure of the environment and determining the vehicle's location on the roadway in a comprehensive manner. It employs laser, ultraviolet, visible, or infrared light to produce images of objects. The emitter dispatches a laser beam that reflects off a mirror, which rotates within the cylindrical housing at a velocity of 10 revolutions per minute [12]. Upon reflection from objects, the laser beam returns to the mirror and is redirected towards the receiver, where it is translated into data. This data is conveyed to a computer, which generates a three-dimensional map of the surrounding environment. The vehicle utilises this map to prevent collisions with other objects. The precision of this map is articulated in centimetres, owing to the minute wavelength of light, which allows for reflections from various surface types and diminutive objects [101]
Radio detection and ranging	RADAR	This device can estimate the relative speed of an object and a vehicle by utilising electromagnetic waves. During the measurement process, it emits a signal and then waits for its return. The frequency of the returned signal experiences a slight alteration during mutual movement, known as the Doppler effect, allowing for the prompt calculation of the vehicle's speed in front [84]. In comparison to LIDAR, radar technology uses a longer wavelength and lower signal energy; however, it cannot describe the shape of the scanned environment. Additionally, it may struggle to identify non-metallic objects or those with certain shapes. Most radar systems operate at around 77 GHz, and the scanning beam is relatively directional. The radar scans the road ahead of the vehicle at a maximum distance of about 200 m. Some vehicles are fitted with two radars working at different ranges. Radar systems are installed on both the front and rear bumpers of a vehicle. The RADAR system detects the surrounding environment, and the ADS (central computing unit) integrates these findings with those obtained from the LIDAR system. The radar system is used for various functions, including detecting oncoming vehicles, their speed, other obstacles, self-parking, blind spot detection, lane change assistance, adaptive cruise control, side collision warnings, and cross-traffic alerts, among others [25]

(continued)

Table 2.2 (continued)

Name	Abbreviation	Description
Ultrasonic sensors		The sensors are strategically mounted on various sides of the vehicle to detect objects in close proximity and to assess the positions of other vehicles during parking manoeuvres. These sensors offer a range of functionalities including parking assistance, collision warning, and lane departure alerts, among others
Video recording devices		The devices are positioned at the apex of the windshield, adjacent to the rear-view mirror, and they generate three-dimensional representations of the roadway in real time. These devices are employed to identify traffic signals, road signs, and unforeseen objects, such as animals or pedestrians, among others. Furthermore, they are capable of recognising various road signs, including 'STOP' signs, pedestrian crossings, and informational boards. Additionally, video cameras assist in recognising specific gestures that other sensors may not interpret, such as waving hands or traffic cones [101]
Global positioning system	GPS	The Global Positioning System (GPS) is a space-based satellite navigation system that provides current location and temporal information anywhere on Earth, provided there is an unobstructed line of sight to four or more satellites. This system serves as the foundation for all maps utilised by vehicles during operation. All satellites transmit signals on two specific frequencies, 1.57542 GHz and 1.2276 GHz; GPS employs these satellites to ascertain the vehicle's present position. Furthermore, GPS maintains the vehicle's course with an accuracy of up to 30 cm. By utilising GPS, a comprehensive map of the region is uploaded to a central computing unit. Additionally, GPS integrates with other systems to ascertain the complete position [101]
Inertial measurement unit	IMU	Data derived solely from GPS is less accurate. Consequently, this data is integrated with the output from the inertial measurement unit (IMU), which utilises a combination of accelerometers, gyroscopes, and magnetometers. The IMU is an electronic device that measures and provides information regarding the vehicle's speed, orientation, and gravitational forces, among other parameters. It assists the GPS in functioning effectively when signals are compromised, such as in tunnels, during adverse weather conditions, or in the presence of electromagnetic interference [101]
Computer	CPU/ADS (autonomous driving system)	Data obtained from each installed sensor is transmitted to a central processing unit, which processes the information at high speed. Utilising sophisticated software, it makes the necessary decisions and communicates the output to electromechanical units, including automatic steering, throttle, and brake systems. This computing system is also connected to the Internet and GPS, thereby facilitating real-time monitoring and updates [71]. This terminology should be referred to as ADS

2.7 Programming Algorithms for Autonomous Vehicles

Autonomous driving systems can execute driving tasks without human intervention by utilising perception, decision-making, and electronic control technologies [7]. These algorithms utilise extensive datasets that gather information about the vehicle's driving history, along with other data, as well as the profiles of other vehicles operating within the same driving environment. In the context of automation levels 3, 4, and 5, where the system assumes complete control over the vehicle (at least for certain durations), the software is responsible for making decisions regarding 'life and death' [3], thereby establishing a sphere of responsibility for incidents related to the movement of the autonomous vehicle.

In the realm of travel planning, three categories of algorithms can be identified, with their sequence indicating a hierarchical structure: (1) mission planning, (2) manoeuvre planning, and (3) path/movement planning [70]. Mission planning constitutes the most abstract and elevated tier of this hierarchy. It utilises road network information to formulate a graph that delineates the path an autonomous vehicle is intended to follow.

The manoeuvre planner resides in the middle layer of the hierarchy and is responsible for calculating the actual driving manoeuvres necessary to reach the waypoints established by the mission planning algorithm. This process is based on the identification of driving contexts, which involve making decisions related to lane changes, road priorities, and safety at intersections, as well as yielding protocols. According to Urmson et al. [94], the subsystem is provided with a cost graph that delineates the expenses associated with various paths leading to the destination, as generated by the mission planner. The cost graph serves as the reference point for current decision-making regarding roadway usage and lane selection by the behaviour subsystem.

Path and motion planning involves computing the actual movement trajectory, effectively avoiding obstacles or collisions. This process can be categorised into two distinct components based on the vehicle's surroundings: (1) Lane and road planning; (2) Zone and parking planning.

In the initial instance, as stated by Ferguson [22], a strategy is devised to adhere to the existing lane and to circumvent both static and dynamic obstacles. To realise this, a trajectory generation algorithm [32] is employed to compute dynamically feasible trajectories. Subsequently, the zone/parking planning algorithm investigates the vehicle's position and orientation, utilising the search algorithm delineated by Likhachev et al. [59].

One of the main challenges facing the development of automated vehicle technology is programming vehicles to handle the complex and dynamic nature of road driving [4]. Automated vehicle technology utilises machine learning techniques [83] that allow an algorithm to predict or decide on 'something' in the external environment without requiring human intervention [14, 63].

The algorithm is meticulously designed to achieve this objective by learning from data, utilising examples, and drawing on its own trial-and-error experiences. Based on this accumulated knowledge, the algorithm independently formulates its own internal decision-making logic regarding the most effective execution of the task [78]. After establishing this internal decision framework, the algorithm is equipped to autonomously make driving decisions in a dynamic driving environment, free from human intervention. For instance, during the 'obstacle detection' phase, the algorithm initiates the process of 'learning' the characteristics of pedestrian appearances by systematically analysing camera footage. A parallel approach would be employed to train the driving system to anticipate pedestrian behaviours [39].

An illustration of machine learning techniques employed during the route planning phase of an autonomous vehicle entails providing the Automated Driving System (ADS) with input data that demonstrates how non-autonomous (human-driven) vehicles negotiate turns, depending on whether the vehicle approaches an intersection with a green light or executes a turn from a stationary position [39]. Consequently, the algorithm will acquire the decision logic relevant to its approach to left turns in the future, aiming to optimise smooth driving and enhance passenger comfort. A critical issue associated with machine learning techniques for assigning responsibility lies in the inability of autonomous vehicle manufacturers to fully anticipate the future behaviour of the ADS [14, 63].

This phenomenon can be attributed to the efficacy of the algorithms, particularly due to their capacity to learn from their experiences and interactions with the driving environment [15]. However, during the programming phase of these algorithms, it is feasible to establish preferences concerning the protected assets of the decisions made in the event of decision-making dilemmas. The programmer delineates the overarching priorities according to which decisions should be rendered and possesses the capability to modify the weighting of the factors influencing these decisions. For instance, Google has incorporated into the decision-making framework of the automated vehicle algorithm the premise that buses and large vehicles are less likely to yield to a vehicle entering the traffic compared to other vehicle types [28, 46].

Nevertheless, the elements requiring rectification or adaptation within an algorithm may not always be clear or predictable until an error arises during the testing phase. Errors are an unavoidable aspect of learning algorithms [63]. Consequently, algorithm testing is one of the most critical components of automated vehicle development, as it clarifies for manufacturers the necessary improvements for their automated driving systems [1]. It is impractical to comprehensively test vehicles against an infinite variety of unforeseen circumstances that may occur within the driving environment [50]. Hence, autonomous vehicles will inherently operate with a certain degree of accident risk [1].

In deliberations concerning criminal liability, it is crucial to consider the decision-making processes of an autonomous vehicle, particularly in situations where the Automated Driving System (ADS) has taken control from the human driver. In extreme scenarios requiring swift reactions, human drivers often act unpredictably, guided by instinct, driving skill, and their physical and mental readiness. In contrast, the decisions made by the ADS in similar circumstances should be viewed as deliberate and well-considered. The basis of these decisions lies in algorithms, specifically the methodology of their programming. These algorithms enable a comprehensive analysis of data gathered during the machine learning process. Consequently, it is vital to assert that the decisions executed by the ADS in road situations, especially those involving ethical dilemmas that require the 'sacrifice' of one value for another, should indeed be characterised as well-considered decisions.

The way algorithms are programmed may consequently influence the preferential treatment of certain road users [61] over others, thereby shaping the risks associated with the behaviour of an autonomous vehicle on the roadway. The rationale behind this preferential treatment relates to the moral implications of the decisions made. It is imperative to engage in a discourse regarding the distribution of risk that is morally acceptable in relation to the decisions made by Automated Driving Systems (ADS) regarding the operation of an autonomous vehicle [56]. Subsequently, it would be the responsibility of lawmakers to regulate the criteria governing decision-making through algorithms [23]. Furthermore, the violation of established algorithms by a driver should be considered an independent prohibited offence.

2.8 Conclusions

Artificial intelligence should be understood as a collective term encompassing various algorithm-based technologies, each distinguished by unique characteristics and intentions for diverse applications. It has been defined in relation to its perceived intelligence, its ability to act autonomously, and its capacity for unpredictable behavioural evolution. Learning is a requisite condition for the advancement of artificial intelligence. The field employs two primary methodologies of 'learning': machine learning and knowledge representation and reasoning. The concept of an artificial agent signifies the connection point between an autonomous vehicle and artificial intelligence. An artificial agent is an information system designed for learning, with the capability to exist independently. In this context, it is referred to as a 'software agent', or it may function as a control element within a device, collectively known as a 'robot'. Thus, the concept of an artificial agent serves as a collective term that encompasses both a software agent and a robot.

Vehicle autonomy, defined as the ability to move independently of human intervention, arises from the use of a control module for assistance systems, commonly referred to as an Automated Driving System (ADS). This control module aligns with the characteristics typical of an artificial agent. The ADS holds internal significance, as it pertains to the vehicle itself, enabling operation without human involvement. Furthermore, it has external significance by impacting other autonomous vehicles, as it facilitates interconnections among them. Within Regulation 2019/2144, the European Union legislator provided definitions for the terms 'automated vehicle' and 'fully automated vehicle'. The former denotes a motor vehicle engineered and constructed to operate autonomously for a limited duration without continuous supervision by the driver, although driver intervention may still be necessary. Conversely, the latter concept describes a motor vehicle designed and constructed to function autonomously without any oversight from the driver. A draft amendment to the

Convention on Road Traffic, prepared by the Global Road Safety Forum of the United Nations Economic Commission for Europe (1960), introduced two novel definitions for 'Automated Driving System' and 'Dynamic Control'. Moreover, Article 65k of the Polish Road Traffic Act defines an 'autonomous vehicle' as a motor vehicle outfitted with systems capable of controlling its movement, enabling it to operate independently of driver intervention while still allowing the driver to assume control at any given moment.

For further consideration, the concept of an autonomous vehicle is based on the following features: (1) a vehicle that is a mechanical means of transport moved by natural forces; (2) equipped with ADS that enables the integration and management of assistance systems installed within the vehicle; (3) resulting in the capability to steer the vehicle without human intervention, both from within the vehicle and externally (internal significance), and to influence other autonomous vehicles (e.g., by forming a column of such vehicles) (external significance).

The Society of Automotive Engineers [83] has delineated six levels of vehicle automation that continuously execute all Dynamic Driving Tasks (DDTs). Vehicles classified under automation levels 3–5 can be regarded as autonomous. Nevertheless, a vehicle may contain a driving automation system that is proficient in providing various driving automation functions operating at distinct levels; however, the level of driving automation exhibited in a specific instance depends on the activated functions.

Level 5 vehicles represent the highest level of automation, in which the Automated Driving System (ADS) can perform the entire dynamic driving task in any scenario, on any road, and under all weather conditions. To operate independently and ensure safety, such vehicles gather extensive amounts of data. This automation level best exemplifies the concept of a self-driving (autonomous) vehicle, where, in specific instances, there may not even be an individual present within the vehicle. Consequently, this degree of automation, referred to as Level 5, is also called 'full automation'. Autonomous vehicles are equipped with various sensors that collect data about the surrounding environment. The algorithms utilise substantial data sets that compile information regarding the vehicle's driving history and the operations of other vehicles within the same driving system. The algorithm strives to achieve this by learning from data, through examples, and via its trial-and-error experiences. Drawing from this accumulated experience, the algorithm improves its internal decision-making logic to achieve optimal task performance. Once this logic is established, the algorithm can independently make driving decisions within dynamic environments, free from human intervention.

It is infeasible to conduct comprehensive testing of vehicles against an infinite array of unforeseen circumstances that may arise within the driving environment. In extreme situations necessitating prompt reactions in traffic scenarios, the determinations made by Autonomous Driving Systems (ADS) are calculated, even when they require the "sacrifice" of one value for another. The basis of these determinations lies within the algorithms, particularly in their programming. This may consequently lead to preferential treatment of certain road users, thereby impacting the risks associated with the autonomous vehicle's behaviour on the roadway. The reasoning behind this preferential treatment stems from the ethical implications of the choices made. There exists an urgent necessity for discourse regarding what allocation of risk is ethically permissible concerning the decisions made by ADS in relation to the movement of autonomous vehicles. In this context, it is imperative for legislators to formulate regulations that govern the criteria for decision-making based on algorithms.

References

1. Ackerman, E.: Toyota's Gill Pratt on Self-Driving Cars and the Reality of Full Autonomy. IEEE Spectrum (2017). https://spectrum.ieee.org/toyota-gill-pratt-on-the-reality-of-full-autonomy. Accessed 21 Feb 2025
2. Act of 6 June 1997. Penal Code (i.e. Journal of Laws of 2025, item 383)
3. Altunyaldiz, Z.: Legal aspects of „autonomous" vehicles. Committee on Legal Affairs and Human Rights, Draft resolution and draft recommendation unanimously adopted by the committee on 9.09.2020. http://www.assembly.coe.int/LifeRay/JUR/Pdf/DocsAndDecs/2020/AS-JUR-2020-20-EN.pdf. Accessed 21 Feb 2025
4. Anderson, M., Kalra, N., Stanley, K.D., Sorensen, P., Samaras, C., Oluwatolala, T.A.: Autonomous Vehicle Technology: A Guide for Policymakers. Research Report, RAND Corporation, Santa Monica (2016)
5. Audi: Audi at NIPS: New Approaches to AI on the Way to Autonomous Driving—Audi UK Newsroom. https://press.audi.co.uk/en-gb/releases/61#:~:text=The%20new%20Audi%20A8%20is%20the%20first%20car,market%20allow%20it%20and%20the%20driver%20selects%20it. i BMW, www.2025ad.com/latest/bmw-driverless-cars-strategy/?WT.tsrc=Newsletter&WT.mc_id=100/2018/ext. Accessed 21 May 2025
6. Automated and Electric Vehicles Act: https://www.legislation.gov.uk/ukpga/2018/18/contents/enacted. Accessed 21 Feb 2025 (2018)
7. Baby, T., Ippoliti, H.Ş., Wintersberger, P., Zhang, Y., Yoon, S.H., Lee, J., Lee, S.C.: Development and classification of autonomous vehicle's ambiguous driving scenario. Accid. Anal. Prev. **200**, Article 107501 (2024)
8. Benenson, R., Petti, S., Fraichard, T., Parent, M.: Towards urban driverless vehicles. Int. J. Veh. Auton. Syst. **6**(1–2) (2008)
9. Bolander, T.: Self-Reference [w:] Stanford Encyclopedia of Philosophy. CSLI, Stanford University. https://plato.stanford.edu/entries/self-reference/. Accessed 22 Feb 2025
10. Bostrom, N.: Superintelligence: Paths, Dangers, Strategies. Oxford (2016)
11. Brachman, R.J., Levesque, H.: Knowledge Representation and Reasoning, San Francisco (2004). https://www.cin.ufpe.br/~mtcfa/files/in1122/Knowledge%20Representation%20and%20Reasoning.pdf. Accessed 21 Feb 2025
12. Chang, Y.-P., Liu, Ch.-N., Pei, Z., Lee, S.-M., Lai, Y.-K., Han, P., Shin, H.-K., Cheng, W.-H.: New scheme of LiDAR-embedded smart laser headlight for autonomous vehicles. Opt. Express **27**(20), 1481–1482 (2019)
13. Proposal for a Regulation of the European Parliament and of the Council Laying Down Harmonised Rules On Artificial Intelligence (Artificial Intelligence Act) And Amending Certain Union Legislative Acts Com(2021) 206 Final). https://www.europarl.europa.eu/RegData/docs_autres_institutions/commission_europeenne/com/2021/0206/COM_COM(2021)0206_EN.pdf. Accessed 21 Mar 2025
14. Copeland, M.: What's the Difference Between Artificial Intelligence, Machine Learning, and Deep Learning? Nvidia, 29 Jul 2016. https://blogs.nvidia.com/blog/whats-difference-artificial-intelligence-machine-learning-deep-learning-ai/. Accessed 21 Feb 2025
15. Dahiyat, E.A.R.: Intelligent agents and liability: is it a doctrinal problem or merely a problem of explanation? Artif. Intell. Law **18** (2010)

16. Deahl, D.: Volvo's Self-driving Cars Are Having Trouble Recognizing Kangaroos. The Verge, 3 Jul 2017. https://www.theverge.com/2017/7/3/15916076/volvo-self-driving-cars-trouble-recognizing-kangaroos. Accessed 22 Jan 2025
17. Definition of Necessary Vehicle and Infrastructure Systems for Automated Driving. https://trimis.ec.europa.eu/project/definition-necessary-vehicle-and-infrastructure-systems-automated-driving. Accessed 21 Feb 2025
18. Erhardt, J., Mona, M.: Rechtsperson Roboter—Philosophische Grundlagen für den rechtlichen Umgang mit künstlicher Intelligenz [w:] Intelligente Agenten und das Recht, Hrsg. S. Gless, K. Seelmann, Baden-Baden (2016)
19. European Commission: COM 2019/168 Building Trust in Human-Centric Artificial Intelligence (2019). https://www.eur-lex.europa.eu/legal-content/EN/TXT/?uri=CELEX:52019DC0168. Accessed 2 Apr 2025
20. Expert Group on Liability and New Technologies New Technologies Formation, Liability for Artificial Intelligence and Other Emerging Digital Technologies (2019). https://www.europarl.europa.eu/meetdocs/2014_2019/plmrep/COMMITTEES/JURI/DV/2020/01-09/AI-report_EN.pdf. Accessed 21 Mar 2025
21. Fagnant, D.J., Kockelman, K.: Preparing a nation for autonomous vehicles: opportunities, barriers and policy recommendations. Transp. Res. Part. A: Policy Pract. **77**, 167–181 (2015). https://doi.org/10.1016/j.tra.2015.04.003. Accessed 21 Feb 2025
22. Ferguson D., Howard T.M., Likhachev M., Motion Planning in Urban Environments: Part II, 2008 IEEE/RSJ International Conference on Intelligent Robots and Systems.
23. Fleetwood, J.: Public health, ethics, and autonomous vehicles. Am. J. Public Health **107**(4) (2017)
24. Gasser, U., Almeida, V.A.F.: A layered model for AI governance. IEEE Internet Comput. **21** (2017)
25. Gerstmair, M., Melzer, A., Onic, A., Huemer, M.: On the safe road toward autonomous driving phase noise monitoring in radar sensors for functional safety compliance. IEEE Signal Process. Mag. **36**(5), 60–61 (2019)
26. Gibbs, S.: What's it Like to Drive with Tesla's Autopilot and How Does it Work? The Guardian, 1 Jul 2016. https://www.theguardian.com/technology/2016/jul/01/tesla-autopilot-model-s-crash-how-does-it-work. Accessed 29 Mar 2025
27. Glancy, D.J.: Privacy in autonomous vehicles. Santa Clara L. Rev. **52**, 1189 (2012)
28. Google: Google Self-Driving Car Project. Monthly Report: February 2016. https://www.driverlesstransportation.com/media/google.com/en/selfdrivingcar/files/reports/report-0516.pdf. Accessed 21 Apr 2025
29. Gruyer, D., Belaroussi, R., Revilloud, R.: Accurate lateral positioning from map data and road marking detection. Expert Syst. Appl. **43** (2016)
30. Guarino, A.: Autonomous intelligent agents in cyber offence. In: Podins, K., Stinissen, J., Maybaum, M. (eds.) 2013 5th International Conference on Cyber Conflict. NATO CCD COE Publications, Tallinn (2013)
31. Hallevy, G.: The criminal liability of artificial intelligence entities—from science fiction to legal social control. Akron Intell. Property J. **4**(2), 171 (2010)
32. Haward, T.M., Knepper, R.A., Kelly, A.: Constrained optimization path following of wheeled robots in natural terrain. In: Khatib, O., Kumar, V., Rus, D. (eds.) Experimental Robotics. The 10th International Symposium on Experimental Robotics. Berlin (2008)
33. Hodula, M.: Criminal liability of the selfdriving vehicles. In: Szakács, A., Hlinka, T. (eds.) Kybernetická kriminalita v čase krízy, Zborník príspevkov z medzinárodnej vedeckej online konferencie Bratislavské Právnické Fórum 2021. Bratislava (2021)
34. Huang, A., Moore, D., Antone, M., Olson, E., Teller, S.: Finding multiple lanes in urban road networks with vision and Lidar. Auton. Robots **26**(2) (2009)
35. IBM: https://www.ibm.com/topics/computer-vision. Accessed 21 Feb 2025
36. Ignatious, H.A., Sayed, H.E., Khan, M.: An overview of sensors in autonomous vehicles. Procedia Comput. Sci. **198** (2022)

37. Ilková, V., Ilka, A.: Legal aspects of autonomous vehicles—an overview. In: 21st International Conference on Process Control (PC), Štrbské Pleso, Slovakia, 6–9 June 2017. https://folk.ntnu.no/skoge/prost/proceedings/process-control-slovakia-2017/data/abstracts/077.html. Accessed 21 Feb 2025
38. ISO/IEC 22989:2022: Information Technology, Artificial intelligence, Artificial Intelligence Concepts and Terminology. https://www.iso.org/obp/ui/en/#iso:std:iso-iec:22989:ed-1:v1:en. Accessed 21 May 2025
39. Jaffe, E.: The First Look at How Google's Self-Driving Car Handles City Streets. CityLab. http://www.citylab.com/tech/2014/04/first-look-how-googles-self-driving-car-handles-city-streets/8977/. Accessed 21 Apr 2025
40. Julier, S., Durrant-Whyte, H.: On the role of process models in autonomous land vehicle navigation systems. IEEE Trans. Robot. **19**(1) (2003)
41. Kajava, K., Torres, A.P.G., Rannisto, A., Sakai, Sh.: Justifying AI regulation: examining multi-stakeholder responses to the AI Act. Telematics Informatics **99**, 102278 (2025). https://doi.org/10.1016/j.tele.2025.102278. Accessed 21 Feb 2025
42. Kant, I.: The Critique of Practical Reason (W.S. Pluhar, translator). Hackett Publishing Company, Inc., Cambridge (2002)
43. Katrakazas, C., Quddus, M., Chen, W.H., Deka, L.: Real-time motion planning methods for autonomous on-road driving. State-of-the-art and future research directions. Transp. Res. Part C: Emerg. Technol. **60** (2015)
44. Kaur, K., Rampersad, G.: Trust in driverless cars: investigating key factors influencing the adoption of driverless cars. J. Eng. Technol. Manag. **48** (2018)
45. Kennedy, C.M.: Reducing indifference: steps towards autonomous agents with human concerns. In: Convention of the Society for Artificial Intelligence and Simulated Behaviour (AISB 2000), Symposium on "AI, Ethics and (Quasi-) Human Rights", Birmingham, 2000, p. 2
46. Kerstetter, J.: Daily Report: Tesla Driver Dies in 'Autopilot' Accident. New York Times, 1 Jul 2016. https://www.nytimes.com/2016/07/02/technology/daily-report-tesla-driver-dies-in-autopilot-accident.html. Accessed 21 Feb 2025
47. Khan, F., Kumar, R.L., Kadry, S., Nam, Y., Meqdad, M.N.: Autonomous vehicles: a study of implementation and security. Int. J. Electr. Comput. Eng. **11** (2021)
48. Kingston, C.K.J.: Artificial intelligence and legal liability. In: Bramer, M., Petridis, M. (eds.) Research and Development in Intelligent Systems XXXIII, Cham (2016)
49. Kirn, S., Mueller-Hengstenberg, C.D.: Intelligente (Software-) Agenten: Eine neue Herausforderung für die Gesellschaft und unser Rechtssystem? FZID Discussion Paper 86, p. 4 (2014)
50. Koopman, P., Wagner, M.: Challenges in autonomous vehicle testing and validation. SAE Int. J. Transp. Saf. **4**(1), 16 (2016)
51. Kose, U., Vasant, P.: Fading Intelligence Theory: A Theory on Keeping Artificial Intelligence Safety for the Future, p. 2. IEEE (2017)
52. Kotoulas, Y.: Mercedes to accept liability for accidents involving its new self-driving tech. Insurance Times, 22 Mar 2022. https://www.insurancetimes.co.uk/news/mercedes-to-accept-liability-for-accidents-involving-its-new-self-driving-tech/1440701.article. Accessed 21 Feb 2025
53. Krasuski, A.: Chmura obliczeniowa. Prawne aspekty zastosowania. Wolters Kluwer (2018)
54. Krasuski, A.: Odpowiedzialność w związku z ruchem autonomicznego pojazdu. Wolters Kluwer, Warszawa (2025)
55. Krasuski: Korzystanie z chmury obliczeniowej przez instytucje finansowe w świetle RODO. Monitor Prawa Bankowego 7–8 (2019)
56. Krügel, S., Uhl, M.: Autonomous vehicles and moral judgments under risk. Transp. Res. Part A Policy Pract. **155** (2022)
57. Legg, S., Hutter, M.: Universal intelligence: a definition of machine intelligence. Minds Mach. **17**(4) (2007)

58. Leminen, S., Rajahonka, M., Wendelin, R., Westerlund, M., Nystrum, A.-G.: Autonomous vehicle solutions and their digital servitization. Technol. Forecast. Soc. Change **185**, 1 (2022)
59. Likhachev, M., Ferguson, D., Gordon, G.J., Stentz, A., Thrun, S.B.: Anytime dynamic A*: an anytime, replanning algorithm. In: ICAPS'2005: Proceedings of the Fifteenth International Conference on International Conference on Automated Planning and Scheduling
60. Lugano, G., Hudák, M., Ivančo, M., Loveček, T.: From the mind to the cloud: personal data in the age of the Internet of Things. In: Zhou, Y., Fischer, M.H. (eds.) AI Love You Developments in Human-Robot Intimate Relationships. Springer (2019)
61. Mamak, K., Glanc, J.: Problems with the prospective connected autonomous vehicles regulation: finding a fair balance versus the instinct for self-preservation. Technol. Soc. **71** (2022)
62. Martinez, R.: Artificial Intelligence: distinguishing between types & definitions. Nevada Law J. **19**(3) (2019)
63. Matthias, A.: The responsibility gap: ascribing responsibility for the actions of learning automata. Ethics Inf. Technol. **6**(3) (2004)
64. Maurer, M., Behringer, R., Furst, S., Thomanek, F., Dickmanns, E.D.: A compact vision system for road vehicle guidance. In: 13th International Conference on Pattern Recognition, vol. 3 (1996)
65. McCarthy, J.J., Minsky, M.L., Rochester, N., Shannon, C.E.: A proposal for the Dartmouth summer research project on artificial intelligence. AI Mag. **27**(4), 12 (1955)
66. Muller, N.J.: Improving network operations with intelligent agents. Int. J. Netw. Manag. **7**(3) (1997)
67. Müller, V.C.: Ethics of artificial intelligence and robotics. In Zalta, E.N. (ed.) Stanford Encyclopedia of Philosophy (Summer 2020; Palo Alto: CSLI, Stanford University), pp. 1–70 (2020). https://plato.stanford.edu/entries/ethics-ai/. Accessed 21 Feb 2025
68. NHTSA: Preliminary Statement of Policy Concerning Automated Vehicles. National Highway Traffic Safety Administration, Washington, DC (2013)
69. Nordström, M.: AI under great uncertainty: implications and decision strategies for public policy. AI Soc. **37**, 1703–1714 (2022). https://doi.org/10.1007/s00146-021-01263-4
70. Okuda, R., Kajiwara, Y., Terashima, K.: A survey of technical trend of ADAS and autonomous driving. Technical Papers of 2014 International Symposium on VLSI Design, Automation and Test, April 2014, 28–30 Apr 2014, Hsinchu, Taiwan. ISBN: 978-1-4799-2776-0
71. Ondruš, J., Kolla, E., Verta, P., Šarić, Ž.: How do autonomous cars work? LOGI 2019—horizons of autonomous mobility in Europe. Transp. Res. Procedia **44** (2020)
72. Paden, B., Čáp, M., Yong, S.Z., Yershov, D., Frazzoli, E.: A survey of motion planning and control techniques for self-driving urban vehicles. IEEE Trans. Intell. Veh. **1**(1) (2016)
73. Pegis, A.C.: Thomas Aquinas, Basic Writings of St. Thomas Aquinas. Hackett Publishing Co. (1997)
74. Pillath, S.: Automated Vehicles in the EU. https://epthinktank.eu/2016/01/11/automated-vehicles-in-the-eu/. Accessed 21 Feb 2025
75. Prasetio, E.A., Nurliyana, C.: Evaluating perceived safety of autonomous vehicle: the influence of privacy and cybersecurity to cognitive and emotional safety. IATSS Res. **47** (2023)
76. Regulation (EU) 2019/2144 of the European Parliament and of the Council of 27 November 2019 on type-approval requirements for motor vehicles and their trailers, and systems, components and separate technical units intended for such vehicles, as regards their general safety and the protection of vehicle occupants and vulnerable road users, amending Regulation (EU) 2018/858 of the European Parliament and of the Council and repealing Regulations (EC) No 78/2009, (EC) No 79/2009 and (EC) No 661/2009 of the European Parliament and of the Council and Commission Regulations (EC) No 631/2009, (EU) No 406/2010, (EU) No 672/2010, (EU) No 1003/2010, (EU) No 1005/2010, (EU) No 1008/2010, (EU) No 1009/2010, (EU) No 19/2011, (EU) No 109/2011, (EU) No 458/2011, (EU) No 65/2012, (EU) No 130/2012, (EU) No 347/2012, (EU) No 351/2012, (EU) No 1230/2012 and (EU) 2015/166 (Text with EEA relevance), PE/82/2019/REV/1, OJ L 325, 16.12.2019, pp. 1–40

77. Regulation (EU) 2024/1689 of the European Parliament and of the Council of 13 June 2024 laying down harmonised rules on artificial intelligence and amending Regulations (EC) No 300/2008, (EU) No 167/2013, (EU) No 168/2013, (EU) 2018/858, (EU) 2018/1139 and (EU) 2019/2144 and Directives 2014/90/EU, (EU) 2016/797 and (EU) 2020/1828 (Artificial Intelligence Act) (Text with EEA relevance), E/24/2024/REV/1, OJ L, 2024/1689, 12.7.2024, ELI: http://data.europa.eu/eli/reg/2024/1689/oj. Accessed 21 Feb 2025
78. Reiley, C.: Deep driving. MIT Technol. Rev. (2016). https://www.technologyreview.com/2016/10/18/156732/deep-driving/. Accessed 21 Apr 2025
79. Revised Amendment proposal to the 1968 Convention on Road Traffic. https://unece.org/fileadmin/DAM/trans/doc/2020/wp1/ECE-TRANS-WP1-2020-1-_Rev1e_.pdf. Accessed 21 Oct 2024
80. Russell, S., Norvig, P.: Artificial Intelligence: Modern Approach. Hoboken (2021)
81. Sadiku, M.N.O., Tembely, M., Musa, S.M.: Internet of vehicles: an introduction. Int. J. Adv. Res. Comput. Sci. Softw. Eng. **8**(1) (2018)
82. SAE International: Taxonomy and Definitions for Terms Related to Driving Automation Systems for On-Road Motor Vehicles J3016_202104. https://www.sae.org/standards/content/j3016_202104. Accessed 21 Feb 2025
83. SAE: https://www.sae.org/standards/content/j3016_202104/. Accessed 21 Feb 2025
84. Šarkan, B., Stopka, O., Gnap, J., Caban, J.: Investigation of exhaust emissions of vehicles with the spark ignition engine within emission control. Transbaltica, Procedia Eng. **187**, 775–776 (2017)
85. Scania 2018: https://www.scania.com/group/en/home/newsroom/news/2018/automated-platooning-step-by-step.html. Accessed 21 Feb 2025
86. Schellekens, M.: Self-driving cars and the chilling effect of liability law. Comput. Law Secur. Rev. **31** (2015)
87. Scherer, M.U.: Regulating artificial intelligence systems: risks, challenges, competencies, and strategies. Harvard J. Law Technol. **29** (2016)
88. Sever, T., Contissa, G.: Automated driving regulations—where are we now? Transp. Res. Interdiscip. Perspect. **24**, 101033 (2024)
89. Shwartz, S., Ben-David, S.: Understanding Machine Learning: From Theory to Algorithms, pp. 448–449. New York (2014); Machová, K.:Strojové učenie. Princípy a Algoritmy, Košice (2002)
90. The Act of 5 January 2011 on Vehicle Drivers (i.e. Journal of Laws of 2024, item 1210, as amended)
91. The Constitution of the Republic of Poland of 2 April 1997 (Journal of Laws No. 78, item 483, as amended)
92. The State of Level 3 Autonomous Driving in 2023: Ready for the Mass Market? Autocrypt (2023). https://autocrypt.io/the-state-of-level-3-autonomous-driving-in-2023/. Accessed 25 Feb 2025
93. Thrun, S., Montemerlo, M., Dahlkamp, H., Stavens, D., Aron, A., Diebel, J., Fong, P., Gale, J., Halpenny, M., Hoffmann, G., Lau, K., Oakley, C., Palatucci, M., Pratt, V., Stang, P., Strohband, S., Dupont, C., Jendrossek, L., Koelen, Ch., Markey, Ch., Rummel, C., van Niekerk, J., Jensen, E., Alessandrini, P., Bradski, G., Davies, B., Ettinger, S., Kaehler, A., Nefian, A., Mahoney, P.: The robot that won the DARPA Grand Challenge. J. Field Robot. **23**(9) (2006)
94. Urmson, Ch., et al.: Tartan Racing: A Multi-modal Approach to the DARPA Urban Challenge, 12 Apr 2007. https://www.grandchallenge.org/grandchallenge/TechPapers/Tartan_Racing.pdf. Accessed 21 Feb 2025
95. Vellinga, N.: From the testing to the deployment of self-driving cars: legal challenges to policymakers on the road ahead. Comput. Law Secur. Rev. **33**(6) (2017)
96. Vojtuš, F., Kordík, M., Dražová, P.: Artificial intelligence and the criminal responsibility—challenges, obstacles and possible solutions. In: 2022 20th International Conference on Emerging eLearning Technologies and Applications (ICETA) (2022)
97. Warren, T.: We Took a Ride in Ford's Self-Driving Car. The Verge, 13 Sept 2016. https://www.theverge.com/2016/9/13/12895690/ford-self-driving-car-ride-detroit-video. Accessed 21 Feb 2025

98. Waymo: https://www.reuters.com/article/usalphabet-autos-self-driving/google-ditched-autopilot-drivingfeature-after-test-user-napped-behind-wheel-idUSKBN1D00MD/?%20il&0,%20and%20Ford:%20www.wired.com/2015/11/ford-self-driving-carplan-google/. Accessed 21 Feb 2025
99. Wei, L.: Legal risk and criminal imputation of weak artificial intelligence. IOP Conf. Ser.: Mater. Sci. Eng. **490**(6) (2019)
100. Yin, J.: Tort liability for damage caused by self-driving cars. In: 2021 International Conference on Public Management and Intelligent Society (PMIS), Shanghai (2021)
101. Yun, H.-S., Kim, T.-H., Park, T.-H.: Speed-bump detection for autonomous vehicles by LiDAR and camera. J. Electr. Eng. Technol. **14**(9), 2155–2156 (2019)
102. Zhai, S., Gao, S., Wang, L., Liu, P.: When both human and machine drivers make mistakes: whom to blame? Transp. Res. Part A **170**, 103637 (2023)

Chapter 3
The Significance of Law Pertaining to the Implementation of Servitisation in Business Models

3.1 Conceptualising the Law in the Context of the Servitisation Process

The research thesis articulated in the introduction to this monograph delineates a connection between legal frameworks and the servitisation of business models, emphasising a transition towards a service-centric paradigm. Legal regulation constitutes a pivotal factor influencing the feasibility of migrating from a product-centric to a service-centric business model. Additionally, a variety of non-legal factors significantly impact this feasibility, including access to appropriate resources, requisite organisational transformations, and the demand for specialised digital mobility services. The concept of adequate resources encompasses a spectrum of elements such as trained personnel, fixed assets, intellectual property, and a network of third-party service providers. The organisational dimension entails adjustments to the internal structures of enterprises, facilitating a shift towards a service-oriented framework. Economic factors relate to the entrepreneurial ecosystem, the capacity to formulate a customer value-in-use proposition, and the provision of dual value. This underscores the complex processes involved in the servitisation phenomenon.

Servitisation is considered successful when the objectives associated with the adopted strategy are effectively achieved. Consequently, servitisation acts as a mechanism for achieving the business objectives outlined in the established strategy for business models. Therefore, legal considerations are crucial for successful implementation.

Benedittini et al. [17] assert that the adoption of a service strategy compels manufacturing companies to engage in a wider variety of activities. This increased involvement heightens exposure to a broader spectrum of regulatory, legal, economic, and technological challenges, in addition to the risks associated with environmental failures. Furthermore, the provision of services induces changes in both competitive and

A. Krasuski, *Legal Challenges of a Transformation from a Product-Centric to a Service-Centric Business Model and Procedures*, Studies in Systems, Decision and Control 610, https://doi.org/10.1007/978-3-031-98901-8_3

operational strategies. For instance, it reveals unconventional competitors, necessitates substantial customer involvement [69, 124], and often relies on business partners for resources, expertise, market insights, and customer access [115].

Consequently, the law may hinder the effective implementation of servitisation or the beginning of the deservitisation process. It is assumed that the business objectives pursued through servitisation comply with legal standards.

The reference point for defining 'law' is the norm of behaviour. According to Oermann et al. [76], in legal theory, some scholars commonly concentrate on the imperative nature of norms. Austin [15] argues that a 'law is a command which obliges a person or persons, and obliges generally to acts or forbearances of a class'. As stated above, a law is a 'command which obliges a person or persons to a course of conduct.'

Other researchers conceptualise a 'norm' as the objective 'meaning of an act which commands or permits, in particular authorises a specific behaviour' [51]. From this perspective, a 'norm' functions as an interpretive scheme for facts [52]. In discussing norms found within normative texts, which are collections of words typically associated with legal norms, it is essential to emphasise the distinction between a "norm' and a "normative sentence' [12, 37]. From this viewpoint, a normative sentence merely serves as a 'linguistic shell' [37] for the norm. This understanding can be generalised to other forms of normative texts: it is crucial to differentiate between normative text and the norm itself. The norm is represented within the normative text and can be discerned through analysing the normative text.

In examining the relationship to behaviour, the sociology of law literature characterises a norm as a 'contradictory stabilised behavioral expectation' [67], indicating its unconditional validity in that 'its validity is experienced regardless of any actual compliance or non-compliance and as it is also institutionalised as such' [67]. Therefore, as indicated by Oermann et al. [76], a norm must exhibit relative stability as it establishes expectations for specific practices.

According to Oermann et al. [76], legal norms embodied in codified normative texts are established by the legislator. This procedure can be understood as a design process undertaken by a creator who drafts the normative text throughout the legislative process. This process is informed by a specific set of pre-existing propositions that include various elements, such as linguistic rules (including semiotics and grammar) and existing legal norms. These propositions and their corresponding elements formulate guidelines for both the norm-setting process and the content of the norms created, which consequently holds significant design guidance.

Consequently, legal norms, reflected in normative texts, may be characterised as cultural artefacts: products of human creation that offer insights into the culture of their creators. Among various purposes, these legal norms aim to facilitate coordinated behaviour within social practices. Furthermore, legal norms may be integrated with a specific control or regulatory framework when the legislator intends to guide social behaviour in a particular direction. This indicates the presence of a discernible intent on the part of the legislator [122], and as demonstrated, legal norms inherently possess an ought-to dimension. Thus, the law encompasses an instrumental component (legal norms with normative content) alongside an action-oriented component (the actual influence of norms on individual behaviour and, consequently, on social practice).

Considering this, it is noteworthy that the term 'law' pertains to the following sources of mandated or proscribed behaviour, occasionally accompanied by the imposition of sanctions.

Firstly, there exist legal norms manifested in legal provisions, wherein these provisions have been publicly disclosed and carry the authority of law. In this context, the legislator delineates all components associated with the content of the legal norm, which encompasses the hypothesis, the disposition, and the sanction, while also specifying the conditions under which the legal provision possesses binding force, as well as its subjective and territorial scope. In this instance, the obligation to act is explicitly articulated within the normative act.

Secondly, the legislator may outline the obligation of relevant (prescribed or prohibited) behaviour in a general manner within a normative act, while requiring the designated public administration body to specify this legal obligation concerning a particular subject. In this scenario, it is the appropriate public administration body that clarifies the legal obligation related to that specific subject in the administrative act.

Thirdly, the legislator may outline the obligation for relevant (either ordered or prohibited) behaviour in a general manner within the normative act, thus requiring the addressee of the legal obligation to specify it on a case-by-case basis, in line with the concretisation criterion established by the legislator. A clear example of such a concretisation criterion is the potential risk of infringement upon the rights or freedoms of natural persons whose personal data are processed, as articulated in Regulation (EU) 2016/679 [80].

The significance of the case law merits emphasis at this juncture. As stated by Oermann et al. [76], in instances of social conflicts that remain unresolved amicably, the necessity for a juridical inquiry arises. In such cases, jurisprudence must facilitate the transition from the instrumental dimension of law to its action-related dimension. Therefore, the entities of jurisprudence encompass legal norms that are articulated or comprehensible within normative texts.

In this perspective of the concept of 'law', there exists a degree of unpredictability regarding the determination of legal obligations when an administrative body acts under the principle of administrative discretion (i.e., issuing a discretionary decision). Administrative discretion should be understood as the authority inherent in a legal norm to select a legal consequence; the disposition of this norm encompasses an alternative, whereby, in a specified state of facts, the authority may opt for one decision among several different options [43]. The scenario differs when the administrative body is required to issue a specific decision within the framework of administrative binding (i.e., issuing a binding decision).

The operation of public administration bodies necessitates the enforcement of the rule of law. For instance, within the Polish legal framework, the principle of the rule of law [121] related to the activities of public administration bodies and the maintenance of legality for the parties and participants involved in the proceedings is articulated in Article 6 and Article 7 ab initio of the Code of Administrative Procedure

[5]. This principle embodies the fundamental tenet of the rule of law as articulated in Article 7 of the Constitution of the Republic of Poland [26, 79]. Furthermore, it is acknowledged in scholarly literature that this principle encompasses different groups of addressees; specifically, the rule of law as outlined in the Code of Administrative Procedure applies within the context of ongoing administrative proceedings governed by that Code [44]. Conversely, the rule of law principle enshrined in Article 7 of the Polish Constitution [26] is universally applicable to all state activities, extending indivisibly to all organs of public power, including legislative, executive, and judicial branches [77].

In accordance with legal statutes and within established parameters, the issuance of both discretionary and mandatory decisions requires a valid legal foundation within substantive law. The absence of such a foundation renders the actions of the public administration body invalid.

The previously described cause of unpredictability can also be understood in relation to the potential risk associated with ineffective implementation of servitisation. This risk may be comprehended as articulated in the literature by Holzer et al. [40], who state that, "fundamental uncertainty, at the time of risk-taking, one cannot know for sure whether the opportunity concerned will be realised; in the worst case, the costs incurred might be greater than any benefit."

According to Zhang and Banerji [119], alongside financial and operational risks, external risks include factors existing outside the organisation that can influence the business landscape. These risks may vary in nature, encompassing changes in technological advancements, regulatory conditions, market trends, globalisation, and capital markets [17].

Khanra et al. [55] assert that manufacturing firms collaborating with third-party service providers often face difficulties in identifying appropriate locations for service centres and ensuring that the quality of service aligns with the firm's established reputation [125]. Moreover, buyer–supplier relationships may encounter challenges when firms that adopt servitisation lack clarity regarding the implementation guidelines [126]. Firms that embrace servitisation might experience similar difficulties in these relationships, which can arise from various factors, including the transparency of information exchange, the strength of operational linkages, changes in the structure of relationships, legal contracts that support relational norms, and improved collaboration in delivering integrated solutions to customers [16].

3.2 Regulation of Disruptive Technology as a Socio-technical Dimension

The literature on servitisation acknowledges its socio-technical dimension, with disruptive developments serving as the reference point (e.g. [42]). In the context of the transformation initiated by servitisation, disruptive innovations such as autonomous vehicles, including those powered by electricity, are driving this change and enhancing environmental sustainability [90].

'Regulation' represents one of the socio-technical dimensions that requires consideration. The literature indicates that the regulatory framework may struggle to keep pace with disruptive innovation [23]. This raises questions about the law's adequacy in effectively regulating technological advancement [99]. Furthermore, regulation may serve as a barrier to experimental solutions that are essential for expediting the low-carbon transformation [14]. Disruptive innovations can evoke serious public policy concerns, leading to a demand for innovative forms of regulation. For instance, emerging 'platform' industries such as Uber and Airbnb have posed challenges to the existing regulatory frameworks governing public transport and hospitality, as well as employment, resulting in regulatory changes within these sectors [46]. A thorough discussion of this topic is provided in Sect. 3.7.

The transition literature argues that if niches are nurtured and supported properly, they can spread and potentially destabilise existing regulations and governance arrangements concerning incumbents [93].

The significance of state legislation and its implementation by public authorities regarding the advancement of digital mobility services warrants consideration within theoretical frameworks. Technological transition, as delineated in the literature [35], is regarded as a substantial technological transformation relevant to fulfilling social functions, including transportation. As a considerable transformation, it includes alterations across several domains, primarily technological change, but also user practices, regulation, industrial networks, and infrastructure.

Given the aforementioned considerations, the emergence and evolution of digital mobility services should be examined from the perspective of a socio-technical transition [114, 120]. Within this transition paradigm, various technologies, numerous socio-economic actors, as well as the roles and networks of these actors, are considered integral components of a cohesive system. This transition is analysed at three distinct levels [35, 36]: landscape (global trend), technological regime (existing system), and niche (emerging innovation). Furthermore, the transition materialises through interactions among these levels. Consistent with this theoretical framework, the landscape of digital mobility services reflects a global trend of digitalisation, which can be viewed as a significant movement within the context of the fourth industrial revolution.

Digital mobility services encompass both technological advancements and innovations in business models [21, 65, 127] and are regarded as niche markets. Within this framework, the prevailing socio-economic conditions relate to mobility services. The favourable conditions conducive to the acceptance of the emerging niche of digital mobility services include ICT infrastructure, mobile usage environments, online services, and platform technologies, among others. Conversely, unfavourable conditions comprise stringent regulations and opposition from existing stakeholders, among others.

The concept of a technological regime enhances the understanding of the situated and pre-structured nature of technical change [89], particularly for third parties. The term 'technological regime' was first introduced in the literature by Nelson et al. [73,

74]. It predominantly pertains to the search heuristics employed by engineers within an industry. According to the authors, 'the sense of potential, of constraints, and of not yet exploited opportunities, implicit in a regime focuses the attention of engineers on certain directions in which progress is possible, and provides strong guidance as to the tactics likely to be fruitful for probing in that direction. In other words, a regime not only defines boundaries, but also trajectories to those boundaries' [73].

The term denotes the technological framework utilised by engineers within an industry and serves as the foundation for their research activities. It has been employed in various ways by scholars of technology [38, 53, 68, 87, 88, 128]. In the study by Rip et al. [88], a technological regime is described as the intricate amalgamation of scientific knowledge, engineering practices, production process technologies, product attributes, user practices, skills and methodologies, as well as institutions and infrastructures that together comprise the entirety of a technology. Conversely, in another study [88], a technological regime is characterised as the grammar or set of rules embedded within the coherent framework of a technology or mode of manufacturing, which organises the search activities of engineers and the policies and actions of various technology stakeholders, including public authorities.

According to Van de Poel [111], various rules may characterise a technological regime. For instance, one can differentiate between rules relating to the design, research and development, production, utilisation, and regulation of a technology. Additionally, rules may also relate to divisions of labour and roles within a technological regime.

In this definition, emphasis is placed on regulations and established practices, which may be perceived as informal regulations or institutions, rather than on the attributes of the knowledge base. These regulations do not exist independently of the technology; instead, they are essential components of it. They encompass formal regulations in the form of compatibility standards, product standards, and regulatory requirements such as emission limit values, alongside accounting rules. Furthermore, they include less formal regulations that are intrinsic to engineering research practices, the division of labour, the organisation of work, management styles, systems, and market conditions, including the degree of competition, barriers to entry and exit, and disadvantages faced by first movers. The significance of the less formal regulations is often greater than that of the formal ones.

A technological regime comprises a cognitive and normative framework alongside a functional structure. This structure includes, among other elements, inter- and intra-company relationships, as well as relationships among technology components, products, processes, skills, and tactics. Both the structure and framework are not static, nor are they entirely coherent. Within a technological regime, one often encounters tensions manifested as product imperfections, capacity constraints, and unmet demands, which serve as focal points for research. These tensions also underpin competition, with interventions from the regime exhibiting distinct patterns of specialisation. The automobile industry serves as an illustrative example; it is characterised as an oligopolistic sector with pronounced forward and backwards linkages, predicated on the mass production of standardised vehicles. In this context, the governing principles differ from those prevalent in sectors such as construction, where technical standards and institutionalised practices are paramount. Instead, the rules in the automobile industry appear to be rooted in organisational rigidities, economies of scale, and prevailing assumptions regarding user requirements concerning vehicle functionality and aesthetics [112].

Innovations emerge within specific niches, shaped by prevailing regimes and their distinct challenges, regulations, and capabilities. These innovations are derived from existing knowledge and resources, tackling the issues inherent in the current systems. Novel technologies are initially conceptualised within established frameworks [34].

The salient aspect of the multi-level perspective is that the continued success of a new technology is influenced not only by processes occurring within the niche, but also by developments at the level of the existing regime and within the sociotechnical landscape. Kemp et al. [54] assert that 'It is the alignment of developments (successful processes within the niche reinforced by changes at regime level and the level of the sociotechnical landscape) which determines if a regime shift will occur.' Changes at the landscape level, for instance, may exert pressure on the regime and create opportunities for new technologies.

In a regime, one can delineate categories of rules, which include core, peripheral, stable, and transient rules. Product and compatibility standards exemplify core rules, whereas supplier–user relationships, encompassing contractual agreements and established interaction patterns, represent transient rules, as a transition to an alternative contractor is always feasible. Design rules and production practices are classified as semi-transient and may undergo gradual modifications. Amending such rules necessitates training, as well as the development and implementation of new procedures. Regulatory standards represent relatively rigid rules, contingent upon the degree to which they are enforced upon companies; however, they do not qualify as core rules for the majority of innovations. They serve as core rules exclusively for compliance innovations. The nature and strength of the rules vary among regimes. Certain regimes, such as the computer regime, are structured around a set of core technologies, whereas in other regimes, the technological aspect is less pronounced. For example, the building regime is predominantly organised around a framework of formal rules, including building regulations, alongside informal rules that are manifested as institutionalised building practices.

As regulation constitutes an integral component of the technological regime, the legislator tasked with formulating laws, along with the appointed administrative bodies responsible for enforcement, significantly influences the capacity of a jurisdiction to facilitate technological advancements. Consequently, when the legislator is dedicated to fostering technological progress, it is imperative that they carefully monitor their regulations to achieve this objective. Amendments to these laws should be enacted well in advance to ensure the effective implementation of technological innovations. The process of revising the law to establish a legal framework for the adoption of emerging technologies must be meticulously planned, considering the intricate nature of this legislative undertaking and accommodating sufficient adaptation periods prior to the enforcement and application of the framework. Furthermore, the legislative process is not the sole area that warrants the legislator's focused attention. The complexity of the regulatory framework encompasses both its material and personal scope, necessitating consideration of the expansive subject matter of the regulation, which includes, but is not limited to, issues such as traffic management, delineation of responsibilities, enforcement mechanisms, and the definition of the rights and obligations of various stakeholders.

It is essential to recognise that establishing a legal framework for the provision of digital mobility services presents a significant challenge for legislators. The legal landscape pertaining to road transport automation, which encompasses product liability, cybersecurity, and intellectual property concerns, is of paramount importance [92]. Furthermore, it is crucial to accurately define the concept of liability by developing substantive distinctions and examining their implications [64]. Among the legislative matters that remain unresolved before autonomous vehicles (AV) can participate independently in traffic is the alignment of their conduct with traffic regulations [78]. Within the realm of traffic regulations, it is necessary to consider issues such as exceptions, conflicts in rules, open texture and ambiguity, alterations in rules, as well as the requirement for common-sense knowledge.

The regulatory framework related to the implementation of digital mobility services must also consider the supervisory responsibilities of regulators. With respect to the utilisation of autonomous vehicles, the existing literature (e.g., [113]) underscores the necessity for regulators to oversee how programmers convert traffic norms into computational code and to establish reliability standards for artificial fact-finding. Specifically, crash algorithms must not infringe upon human dignity and should align with the priorities of the legal order [75].

3.3 Economic Freedom as a Reference Point for the Servitisation Process

The reference point for research concerning the legal challenges of servitisation is the economic freedom enshrined in law, along with the legal grounds for its limitations. The principle of 'economic freedom' occupies a central position in the discourse, particularly regarding the entrepreneur and their capacity to act and effectuate change. The notion of 'entrepreneurship as innovation' has been articulated within the economic literature. This notion aligns with the adoption of digital business models by entrepreneurs in relation to the provision of mobility services. Furthermore, the context of implementing mobility services with respect to autonomous vehicles reinforces the argument concerning the advancement of innovation by entrepreneurs.

According to the concept of 'entrepreneurship as innovation', the entrepreneur is recognised as an innovator (e.g., [95]). The entrepreneur serves as a catalyst for economic change by introducing new products, production methods, markets, sources of supply, or industrial combinations, thereby propelling the economy into the process of creative destruction [94]. Consequently, the process of servitisation exemplifies the transformation indicated by Schumpeter. This process of creation delineates the characteristics of the entrepreneur. However, the management process within the enterprise that introduced the change is of lesser significance [20]. Furthermore, according to Hailemariam et al. [39], economic freedom encompasses essential institutional features such as the protection of property rights, the rule of law, and limited government interference.

A reference to the rule of law necessitates the identification of legal sources that protect economic freedom.

The Charter of Fundamental Rights of the European Union [107] ('Charter') represents the inaugural legally binding document that explicitly acknowledges the freedom to conduct business within the framework of EU law. In accordance with Article 16 of the Charter, this freedom to conduct business, in compliance with Union legislation as well as national laws and practices, is duly recognised.

Usai [110] asserts that the absence of recognition of this right as a fundamental right within both the European Convention on Human Rights (1950) and the Treaties renders this position inconsistent and contrary to the objectives of the European Union legal order. The Court of Justice of the European Union ('CJEU'), the European Commission, and the European Court of Human Rights ('ECHR') [33] have utilised Article 1 of the Additional Protocol to the ECHR, which pertains to the protection of private property, as a foundation for deriving principles that safeguard the right to economic initiative [66].

Freedom is referred to in Article 2 of the Treaty on European Union [108] ('TEU') as one of the 'values on which the Union is founded'. These principles are not merely values of the European Union; they are explicitly stated to be 'common to the Member States' and are designed to demonstrate a commitment, on behalf of the European Union, to the postulates of liberal-democratic constitutionalism [117].

In European Union law, the concept of economic freedom is recognised as a general principle of EU law, a fact substantiated by the existing literature on the matter: 'The general principle of EU law referred to in Article 6 TEU, derived in the case law of the CJEU from the constitutional tradition common to the Member States, as well as the fundamental right guaranteed in Article 16 of the Charter' [31].

In the legal literature, it is emphasised that economic freedom has been infrequently referenced in the case law of the Court of Justice of the European Union ('CJEU') [61]. The CJEU has addressed economic freedom only on a limited number of occasions, which constitutes a relatively small figure considering the extensive influence of European Union law on the domain of economic activity [61]. The literature highlights that the CJEU does not endeavour to define the essence of economic freedom. Furthermore, the literature asserts that 'The dominant perspective here is the assessment of the restriction of such freedom in the aspect of a specific' general interest' that is to implement a given restriction' [19].

It is imperative to underscore that the European Union's institutions' position regarding the freedom of economic activity is fundamentally anchored in the principle of freedom of establishment, alongside the principle of freedom in the provision of services. This assertion is corroborated by the existing literature, which indicates that 'The European Union's attitude to the freedom of economic activity is based on the principle of freedom of establishment and the principle of freedom to provide services. These principles are among the guiding principles of the common market' [59].

The principle of freedom of establishment is enshrined in Article 49 of the Treaty on the Functioning of the European Union (TFEU). This article explicitly prohibits the imposition of restrictions on the freedom of establishment for nationals of any Member State within the territory of another Member State. In general, entities should be treated in accordance with the conditions established in the legislation applicable to their own citizens. The case law of the Court of Justice of the European Union interprets the freedom of establishment in a broadly inclusive manner.

There exists a distinction between the fundamental nature that facilitates the establishment and operation of a business. As indicated in the literature, this pertains to 'the establishment and operation of an enterprise (commencement of business activity) in any place of choice' [60]. Consequently, this matter can be regarded as secondary in nature. According to the literature, this implies that an entity already engaged in business activities in one of the member countries of the community may extend its business operations into the territory of another member country within the European Union. This may encompass the establishment of a new subsidiary or the relocation of part of the company to another country.

In principle, a national of a Member State may provide the services specified in Article 57 of the Treaty on the Functioning of the European Union ('TFEU') under the same conditions as a national of a particular country when delivering a service. Services, as outlined within the framework of the TFEU, primarily include: (a) industrial activities; (b) commercial activities; (c) craft activities; and (d) the practice of liberal professions [41]. Consequently, this broad definition, while detached from the commonly accepted understanding of services as intangible in nature, is fundamentally linked to the notion that the freedom to provide services is subsidiary to the principle of freedom of establishment.

Economic freedom, as defined in both the law of the European Union and the Polish Constitution [26], maintains a consistent meaning. It is recognised as both a systemic principle and a legal framework for freedom. The literature on the subject suggests that the Polish Constitution [26], similar to the primary law of the European Union, allows for the possibility of imposing restrictions on the freedom of economic activity (freedom of establishment), while delineating the criteria for the legitimacy of such restrictions" [59].

Consequently, the relationship between Polish legislation regarding economic freedom serves as a reflection of European Union law in this domain.

Economic freedom comprises various partial freedoms. The scholarly literature on the subject indicates that, in the jurisprudence of the Polish Constitutional Court, attention is often directed towards two specific partial freedoms: the freedom to undertake business activity and the freedom to conduct business activity [104]. Within this framework, further distinctions can be made regarding additional partial freedoms. Consequently, it is generally accepted in the literature that initial entry into the market, specifically the commencement of business activity, is constitutionally protected [104]. The freedom to undertake business activity grants individuals the opportunity to select both the location for conducting business activities and the organisational and legal structure of their business ventures, as well as the type of activities they wish to pursue.

However, freedom is not absolute; it is constrained by provisions concerning the acquisition of various licenses, concessions, permits, consents, and entries in the registers of regulated activities by the entrepreneur. Therefore, the primary role of the state is to avoid establishing unnecessary barriers and, above all, to foster a regulatory environment in which competition protection laws are enacted, thereby facilitating business activities. This notion is underscored by Zaradkiewicz [118] in his assertion that 'decisions in the course of conducting business activity are protected regarding, for example, the choice of the object of activity and its change, the freedom to choose a contractor and the freedom to shape the content of contracts'.

The constitutional issue of economic freedom, as outlined in the Polish Constitution [26], pertains to every individual, particularly every natural person, due to their inherent and inalienable dignity and autonomy.

The term 'everyone' cited previously should encompass individuals currently operating a business, in addition to those aspiring to initiate one. The mere circumstance of an individual not yet being an entrepreneur does not negate the protection afforded by economic freedom. The Polish legislator has instituted various exceptions to the principle, asserting that 'everyone' is a subject of economic freedom.

Article 37(2) of the Polish Constitution [26] stipulates that exceptions to this rule concerning foreigners are established by law. In creating these exceptions, the legislator is required to comply with European Union law. Consequently, Poland is prohibited from discriminating against individuals based on their nationality. Citizens of European Union member states, as well as individuals from other countries who enjoy rights and freedoms concerning the intra-EU market based on specific agreements with the European Union, are granted the status of Polish citizens. This status is also extended to citizens of Ukraine who, due to the armed conflict, have been afforded special privileges in various areas, including employment and entrepreneurship.

The second constitutional exception is delineated in Article 107 of the Polish Constitution [26]. The legislator in this article stipulates that deputies, within the framework specified by the Act, are prohibited from engaging in economic activities intended to obtain any material benefits from the State Treasury or local government, and they are likewise restricted from acquiring such properties. Furthermore, this article of the Polish Constitution [26] also pertains to senators. The prohibition against combining the roles of a deputy or senator with business operations is primarily designed to prevent a scenario in which deputies and senators may make decisions to secure personal financial gains, thereby neglecting the broader interests of the state.

Such exceptions may also be introduced through ordinary legislation. An illustrative example of this restriction is Article 24f(1) of the Act on Local Government [11]. This provision stipulates that councillors are prohibited from engaging in business activities either on their own behalf or on behalf of other entities that utilise municipal property in the municipality where they have received a mandate.

It should be noted that economic freedom is not applicable solely to natural persons. This is due to the ability to choose the manner in which to conduct organisational and legal business activities, as well as the freedom of association, as stipulated in Article 58 of the Polish Constitution [26]. Economic freedom also pertains to other legal entities governed by private law—these include entities with an associating character, as well as commercial law partnerships, cooperatives, associations, capital companies of commercial law, and foundations. It is, however, important to emphasise that freedom is inherently vested only in natural persons, while the aforementioned entities exercise economic freedom solely in an indirect capacity, predicated on the individuals who constitute them.

The essential conditions governing the imposition of restrictions on economic freedom are outlined in the Polish Constitution [26], which stipulates that formal requirements must be observed, particularly regarding the statutory framework and substantive conditions pertinent to the existence of a significant public interest.

Therefore, based on the existing literature on the topic, it is essential to differentiate between restrictions on economic freedom that apply universally to all entrepreneurs and those specific obligations concerning the legalisation of economic activities, all registration requirements pertaining to taxation, insurance, and public statistics, in addition to obligations related to the use of markings (see Judgement of the Constitutional Court: SK 35/08 [47]).

Economic freedom is contingent upon certain limitations. According to Article 22 of the Polish Constitution [26], the right to economic freedom may be restricted only by statute and solely for a significant public interest. Nevertheless, scholarly literature presents a dilemma in this regard, particularly concerning the question of whether any statutory regulation related to the initiation, execution, and termination of business activities—including the definition of prohibitions, directives, and the resultant obligations of entrepreneurs—should be considered constraints on the freedom of economic activity, which may be instituted exclusively for a substantial public interest [59].

The aforementioned inquiry, which is documented in the literature pertaining to the subject, holds considerable significance, as it is posited that not all regulations governing economic activity equate to limitations on economic freedom. Such limitations encompass, among others, obligations associated with the legalisation and registration of business activities, adherence to requirements concerning the labelling of goods, accounting principles, public law fees, and so forth. The differentiation of the norms under discussion is contingent upon the foundational interpretation of the concept of economic freedom. Should we perceive economic freedom as unencumbered liberty, then any form of interference in this domain will represent a greater or lesser limitation.

Consequently, the literature on the subject also addresses the concept of freedom, which inherently includes internal limitations arising from the objective. This objective may encompass the archetype of an entrepreneur, who acts not solely for personal gain but also contributes to the common good through their endeavours. It is important to underscore that, irrespective of the delineation of formative norms which underpin the legal order of the state and should remain unquestioned, such norms impose restrictions on economic freedom. It is noteworthy that, fundamentally, there are ongoing debates regarding whether these restrictions indeed jeopardise economic freedom [102, 103].

Regarding the concept of restricting economic freedom, it is important to note that, in line with the existing literature, 'restrictions on economic freedom should be treated as legal provisions imposing obligations on the entrepreneur that exclude or limit the freedoms related to taking up and conducting business activity' [129]. The introduction of restrictions on business activity necessitates a formal condition, specifically mandatory compliance with statutory requirements, alongside a substantive condition that pertains to the necessity of an essential public interest [102, 103].

Consequently, it may be inferred that, in accordance with the stipulations outlined in the existing literature on the topic, "As for the first condition, one should agree with such a solution. It is difficult to deny the belief that the only way to limit a constitutional principle by a legal act other than a statute. The second condition is nothing more than a vague concept, provided with a general clause, and therefore its understanding may encounter difficulties" [18].

In the discourse surrounding economic freedom, the literature distinguishes between subjective and objective limitations. In this context, subjective restrictions typically manifest as prohibitions directed towards specific groups of individuals who, due to their characteristics, are considered unsuitable to initiate and operate a business [102, 103]. For instance, as noted in Article 24f(1) of the Act on Municipal Self-Government, a restriction may arise in the form of a prohibition preventing the head of the commune from engaging in business activities. Additionally, a personal restriction may also take the form of a judicially mandated ban on conducting business operations. Such a prohibition may stem from bankruptcy proceedings or serve as a punitive measure. This type of ruling may be issued against the debtor, who can be either a natural person or an individual authorised to represent a given entrepreneur, in cases leading to bankruptcy.

Conversely, regarding objective restrictions, they demonstrate greater diversity, as highlighted in the relevant literature. This literature suggests that such restrictions may take the form of objective limitations in the strict sense (i.e., the freedom to select the subject of economic activity), territorial constraints (i.e., the freedom regarding the location of business operations), or temporal limitations (i.e., the freedom to determine the duration of business activities) [59]. Subject restrictions, on the other hand, concern the type and manner of conducting business activities. Furthermore, it is crucial to emphasise the existence of subjective limitations. This type of restriction involves the necessity of commencing a business only after meeting the requisite conditions for obtaining qualifications, completing a course, possessing specific experiences, or acquiring relevant education, among others.

The permissible scope for restricting economic freedom is outlined in Article 22 of the Polish Constitution [26]. This provision states that restrictions on the freedom to initiate and engage in business activities are allowed only through statutory law and solely in pursuit of significant public interest. Consequently, the Polish legislator establishes two paramount conditions for the admissibility of restrictions on economic freedom: formal (exclusively through statute) and substantive (in the context of significant public interest).

The primary prerequisite for the admissibility of a restriction on economic freedom is the formal premise. According to Article 22 of the Constitution of the Republic of Poland, a restriction on the freedom of economic activity may only be imposed by statute. This restriction, based on linguistic interpretation, should not be defined in subordinate acts. Nevertheless, in light of dynamic economic development, both in Poland and globally, it proves exceptionally challenging to limit economic freedom solely through statute. Undoubtedly, the distinction between the statute's content and the executive regulation would be appropriate if the content of the restriction were consistent with Article 92 of the Polish Constitution [26].

The delineation of freedoms as well as personal and economic rights presents a theoretical foundation for articulating restrictions within acts of sub-statutory rank. Nonetheless, concurrently, it has been observed in the existing literature regarding this matter that while the former necessitates more meticulous adherence to both substantive and formal requirements, the latter may be subject to more extensive limitations, both in substance and form.

Another condition necessary for the introduction of restrictions is the 'important public interest'. The very concept of public interest holds significant weight within the realm of administrative law, as it delineates the extent of permissible state interference in the functioning of individuals. This interference is relevant both at the level of legislative authority, which is responsible for formulating substantive administrative law, and at the level of public administration bodies that implement said law [102, 103]. Nevertheless, it is imperative to underscore that due to ongoing changes in the surrounding world, the scope of the concept of public interest remains challenging to define and is in a constant state of flux. Furthermore, public interest is multifaceted, it can serve as a basis for restricting rights or freedoms, while simultaneously acting as a rationale for establishing appropriate mechanisms aimed at safeguarding specific values [123].

It is imperative to underscore that the concept of "important public interest" presented in the relevant literature must be interpreted more expansively than the general criteria for limiting freedoms and rights outlined in Article 31(3) of the Polish Constitution [26]. Szafrański emphasises that this assertion pertains to the correlation between the conditions for the permissibility of restrictions as specified in Articles 22 and 31(3) of the Polish Constitution [26]. The former provision addresses solely the freedom of economic activity, whereas the latter delineates the conditions under which restrictions on freedoms and rights in general may be applied. These conditions include public security, public order, the protection of the environment, public health and morals, as well as the freedoms and rights of others.

Within its scope, "important public interest" encompasses the premises. Article 31(3) of the Polish Constitution [26], alongside other constitutional values not enumerated in the aforementioned article, constitutes this essential public interest. Such an important public interest serves as a basis for imposing greater restrictions on economic freedom than those pertaining to the rights and freedoms delineated in Chapter II of the Polish Constitution [26].

It should also be emphasised that the concept of 'important public interest' is one that undoubtedly evolves over time. Its interpretation may differ significantly today compared to a decade or several years ago, and it is probable that the breadth of this concept will continue to shift in the future. Consequently, what was previously classified as 'important public interest' may no longer hold the same significance today, and conversely, new matters may emerge as deserving of such categorisation.

An analysis of the premises permitting limitations on economic freedom necessitates an examination of the principle of proportionality. According to [97], this principle delineates the boundaries of governmental intervention regarding the rights and freedoms guaranteed to citizens.

The conflict that arises between the public interest and the interests of the individual renders the principle of proportionality an essential component of the legislation within democratic states.

It is therefore imperative to underscore that the principle under discussion has not been explicitly incorporated into the constitutional guarantees of rights and freedoms. Concurrently, this principle has been delineated in Article 31, Section 3, Clause 2 of the Polish Constitution [26]. Furthermore, the essence of the principle of proportionality dictates that public authorities must conduct themselves in a manner that does not impose excessive restrictions on the fundamental rights and freedoms of individuals.

The principle of proportionality, as articulated by the judgment of the Constitutional Tribunal, should be understood as a principle that encompasses:

(1) The principle of usefulness, which involves assessing whether the regulation applied, resulting in limitations on individual rights and freedoms, will achieve the intended purpose.
(2) The principle of necessity, which seeks to determine whether additional regulations are essential for the protection of specific goods; if so, it identifies which of the available protective measures is the least burdensome.
(3) The principle of proportionality sensu stricto, also known as the principle of undue interference, requires verifying whether the benefits accrued from a regulation justify the burdens imposed on individuals concerning the restriction of their rights, as outlined by the Polish Constitutional Court (Judgment K 35/11) [48].

The limitation of economic freedom, beyond satisfying formal prerequisites, necessitates a substantive justification. The evaluation of the legitimacy of interference in economic freedom must be contingent upon the nature of the economic activity, its specific classification, and the extent of legal regulation, which is influenced by public interest [98]. The authority is expected to employ the least restrictive measures in curtailing freedoms and rights.

In light of the aforementioned, limitations on the freedom of economic activity are essential for the effective operation of a democratic state, primarily due to the principle of proportionality. Nevertheless, it is imperative that such limitations do not undermine the fundamental nature of economic freedom. As previously noted, the restriction in question is deemed permissible solely on the basis of significant social interest. Furthermore, it must be implemented in line with the established norms of systemic interpretation and in a complementary and exceptional manner, while also considering the provisions set forth in the previously discussed Polish Constitution [26].

3.4 Assumptions Concerning the Legal Framework for the Provision of Digital Mobility Services

Currently, the legal framework lacks explicit provisions addressing the provision of digital mobility services concerning the use of autonomous vehicles. Considering the principle of economic freedom and the constraints it entails, the legal sources relevant to the provision of digital mobility services should be determined based on the following interconnected assumptions.

The initial assumption suggests that the motor vehicle is experiencing a decline in its status as a symbol, due to the shift in preferences of the new generation from car ownership to car usage. As a result, consumer interest is particularly focused on the primary function of the motor vehicle, which is to enable the transportation of individuals and goods over a specified distance. This forms the second assumption.

The third assumption is based on an examination of the previous two assumptions. Essentially, it involves the exclusion of the motor vehicle from the construction of an individual's image and the evidence of material status that has been achieved. In this context, the legal relationship of the motor vehicle holder with the vehicle as a tangible object loses significance. Instead, the focus shifts to the ability to carry out the tasks of transporting goods and people that can be facilitated by a particular motor vehicle. While exceptions must be acknowledged concerning the desire to own vehicles for collectable purposes, the values articulated above represent an expression of emerging trends in wheeled transportation, particularly in urban environments. Given this perspective, the vehicle is regarded through the lens of the functionalities it offers. Consequently, it becomes feasible to explore specific modes of transport as alternatives to motor vehicles. An illustrative example of this is the emergence of shared e-scooter services, which serve as alternatives to conventional car transportation [1]. The evolving attitude towards the motor vehicle as a mode of transport should be acknowledged as a significant trend. With this understanding, under the third assumption, the scope of services aligned with this trend centres on the transportation function (car-sharing, ride-sharing) and/or on the provision of access to transformative services through digital means.

The fourth assumption is based on the relationship between digital mobility services and motor vehicles, which serve as the primary means of transport. Essentially, the vehicle highlights the importance of providing digital mobility services. The reliance of digital mobility services on the vehicle is evident from the fact that any restrictions or limitations placed on the vehicle negatively affect the provision of related digital mobility services. From a value creation standpoint, the relationship between digital mobility services and the vehicle is bi-directional, leading to a state of interconnectedness. Considering the emerging trend among consumers to prioritise vehicle functions over ownership, it follows that the production of vehicles (equipment) depends on the demand for digital mobility services.

In accordance with the fifth assumption, the reliance on digital mobility services for the vehicle, as a mode of transportation, becomes increasingly significant when the vehicle is considered as an autonomous entity in which the driving function is either assisted or replaced by artificial intelligence. In this context, it is crucial to emphasise that the implementation of autonomous vehicles in public traffic, which varies by different levels of automation, is at various stages across distinct nations. Additionally, some countries have yet to grant official approval for autonomous vehicles classified with the highest levels of automation.

According to the sixth assumption, since autonomous vehicles exemplify disruptive technology, there is a lack of a dedicated legal framework and basis for regulatory action. Attempting to adapt the existing legal structure to accommodate autonomous vehicles through interpretative practices proves challenging and fails to ensure uniformity, especially in the initial stages, until a reputable judicial stance is established. This situation poses a potential threat to the legal security of digital mobility customers and could influence regulatory predictability.

In accordance with the seventh assumption, the integration of artificial intelligence through an automated driving system (ADS) in autonomous vehicles, as part of the product, requires ethical consideration. This ethical stance has significant implications for the development of the legal framework.

According to the eighth assumption, the implementation of autonomous vehicles, which aims to facilitate self-driving transportation on public roads, necessitates modifications to roads designated for use by such vehicles. The infrastructure defining permissible areas for autonomous vehicles will also undergo alterations. Wiseman [116] anticipates that the introduction of fully autonomous vehicles will lead to the elimination of interchange stations and central stations from roadway infrastructure. This assertion is based on the belief that driverless cars will possess the capability to transport passengers from virtually any location to their intended destinations. At the fourth level of automation, the Automated Driving System (ADS) executes all dynamic driving functions, thereby excluding human drivers from engaging in these tasks. Nevertheless, such vehicles operate within specified closed zones that constitute an Operational Design Domain (ODD, for instance, a mining area). Roads situated in designated open zones represent a crucial element of the overarching theme concerning dedicated infrastructure for autonomous vehicles. A significant challenge in this regard lies in integrating ODDS into the interconnectivity of various modes of transportation, including public transport. Autonomous vehicles are currently utilising and will continue to utilise public roads alongside both autonomous and non-autonomous vehicles. The legal challenge involves establishing regulations governing operation, particularly concerning the transfer of non-verbal communication from human drivers to future interactions between artificial agents and human operators. Furthermore, another legal concern related to operating autonomous vehicles includes the issue of civil and criminal liability in the event of traffic incidents, including those resulting in fatal outcomes. Additionally, a vital consideration pertains to the authority governing the operation of autonomous vehicles. The notion of 'dedicated infrastructure for the movement of autonomous vehicles' encompasses not only the physical roadways but also the communication resources necessary for delivering digital mobility services and managing autonomous vehicle operations.

According to the ninth assumption, the production of autonomous vehicles is essential for achieving the objective of digital mobility services, as outlined in the second list. Even though consumer trends emphasise the use of a vehicle's transportation functions rather than viewing the vehicle as an object, it is imperative that individuals purchase these vehicles (including autonomous vehicles) to facilitate transportation. Furthermore, autonomous vehicles must undergo regular inspections that include, in addition to the standard evaluations of vehicle components, assessments related to those components responsible for ensuring autonomous driving capabilities. Additionally, it is necessary for such autonomous vehicles to have adequate insurance coverage against potential accidents while operating on public roadways.

According to the tenth premise, the marketing and provision of digital mobility services, along with the operation of autonomous vehicles, place the customer at the centre of the research. The customer may be categorised as either a legal entity or a natural person. The latter is entitled to special consideration by the legislator under consumer protection legislation. Regardless of whether the customer is a legal entity or a natural individual, the provision of digital mobility services also pertains to a natural person, such as an employee of a customer who is a legal entity. The provision of digital mobility services and the operation of autonomous vehicles require the processing of substantial volumes of data. The processing of this data is subject to the regulations governing electronic communication, and when it involves personal data, it is further regulated by data protection legislation, including Regulation (EU) 2016/679 [80].

3.5 Legal Regulations Pertaining to the Provision of Digital Mobility Services

The analysis of the sources of law applicable to the provision of digital mobility services within the framework of autonomous vehicle utilisation necessitates the identification of the following research directions.

Initially, it is necessary to ascertain whether the relevant legal framework directly regulates the provision of digital mobility services through the applicable legal provisions.

Secondly, suppose the provision of digital mobility services is directly regulated by the pertinent legal framework. In that case, it becomes necessary to ascertain whether this legislation imposes restrictions regarding the provision of such services. This inquiry should also encompass the acceptability of utilising autonomous vehicles, which serve as a benchmark for the delivery of these services.

Thirdly, in instances where the relevant provisions do not explicitly govern the use of autonomous vehicles in economic transactions, or do so inadequately, it is essential to determine whether the use of autonomous vehicles and the provision of associated digital mobility services is permissible through an interpretation of the applicable provisions.

Fourthly, it is essential to ascertain the degree of complexity inherent in the legal system of a particular jurisdiction, and consequently, the potential for rapid changes to the law. Relevant sources of law should be identified within the specified areas of regulation, based on the assumptions articulated in Sect. 3.4. Accordingly, the regulatory areas should be delineated distinctly for the two subjects, namely 'digital mobility services' and "autonomous vehicles," while considering their interdependence as described in Sect. 3.3.

With respect to 'digital mobility services', as defined in Chap. 4, the legal framework could be established based on distinct regulatory areas:

(1) Eligibility to conduct business activities comprising the provision of a single digital mobility service or a suite of digital mobility services;
(2) Legal requirements pertaining to providers of business mobility services;
(3) Legal requirements concerning the resources employed in the provision of business mobility services;
(4) Legal requirements regarding the relationships with customers of digital mobility services;
(5) Legal status of customers of digital mobility services and the scope of customer rights related to the provision of these services.

The subsequent sections elaborate on the areas of regulation with comprehensive detail.

The initial area of regulation concerns the determination of whether, under the applicable legal framework, a restriction on the provision of a specific digital mobility service has been enacted. This involves establishing subject-matter prohibitions regarding the offering of a particular type of digital mobility service. If such subject-matter prohibitions are in place, the provision of a digital mobility service becomes untenable. The regulatory challenges related to the provision of taxi-hailing services in South Korea are addressed in Sect. 3.7 below. As outlined in this section, to evaluate the permanence of the prohibition on the provision of certain digital mobility services, it is essential to ascertain its origin; specifically, whether it arises from a legal norm stipulated within a provision that directly pertains to the provision of digital mobility services, or if the provision containing the aforementioned legal norm does not directly relate to these services, and its application has been extended to digital mobility services by virtue of legal interpretation. These considerations regarding the nature of legal sources are crucial for assessing the permanence of this prohibition.

The regulatory areas delineated in points 2 through 6 become pertinent if, based on the initial area of regulation, a prohibition on the provision of digital mobility services is not established. Such a prohibition may not be directly enshrined in law. However, the legislator may stipulate the requirements for obtaining administrative authorisations, upon which the admissibility of providing digital mobility services (a licensing regime) depends. In this context, the legal classification of the specific digital mobility services to be provided is critical, along with the identification of the type of licence and the procedure for obtaining it. Regarding the issues associated with the licensing process for the provision of such services, it is essential to establish the subjective conditions for the applicant seeking a licence, specifically addressing the question of who may apply for such a licence. The second regulatory area is dedicated to the subjective conditions.

The second area of regulation relates to legal requirements associated with providers of business mobility services. These requirements are subjective. Consequently, it is essential to ascertain whether the provision of specific services is permissible within a particular jurisdiction when the provider has certain characteristics. Therefore, it is feasible to regulate the obligation to deliver services as a legal entity registered in the country where these services are intended to be provided. The rationale for implementing such a requirement may be to enhance the efficiency of law enforcement within the relevant jurisdiction.

Nevertheless, the Charter of Fundamental Rights [109] articulates in Article 16 the right to engage in business activities under both Union law and national law, as well as applicable practices (refer to Sect. 3.3). In this realm of regulation, it is imperative to establish requirements concerning the legal structure relevant to the execution of this particular economic activity. Primarily, it must be ascertained whether this activity can be conducted as an individual enterprise or if it necessitates the establishment of a legal entity specifically designated for this type of economic activity. For instance, in Poland, certain legal structures have been stipulated for specific types of economic activities. According to Article 6, paragraph 1 of the Act on Insurance and Reinsurance Activities [4], an insurance company is permitted to engage in insurance activities solely in the form of a joint-stock company, a mutual insurance company, or a European company as delineated in Council Regulation (EC) No 2157/2001 [25]. Additionally, it is essential to determine whether a given jurisdiction has imposed restrictions that may pertain to the personal composition of the governing bodies of a company engaged in economic activities related to the provision of these services.

In light of the various categories of digital mobility services outlined in Chap. 4, which pertain to the implementation of autonomous vehicles, the third regulatory area emphasises connectivity as a crucial component in providing such services. First and foremost, connectivity must be understood within the context of internet accessibility, serving as a means of delivering digital mobility services through specialised applications while facilitating transportation agreements with digital platforms acting as intermediaries. Section 3.6 elaborates on the legal sources pertinent to the provision of digital mobility services, conceptualised as e-commerce operations. Furthermore, connectivity is essential for the operation of autonomous vehicles at both partial and full levels of driving automation. Additionally, connectivity is necessary for the coordinated behaviour of autonomous vehicles, specifically in executing tasks intended for a collective of artificial agents in the form of a platoon or swarm.

This resource can be obtained through collaboration with third parties, such as providers of electronic communications networks and services. In this context, it is essential to ascertain the purpose of this cooperation and the associated role of the provider of digital mobility services. Will it act solely as an intermediary between the client of the digital mobility service and the provider of electronic communication services, or will it acquire these services on its own behalf to subsequently resell them to the client? The legal relations between these entities are examined in Chap. 4.

The fourth area of regulation pertains to the legal requirements associated with the aspects related to the relationship between providers of digital mobility services and their customers: (1) the content of the offer for the provision of digital mobility services and the method for delivering such services; (2) the content of the agreement for the provision of digital mobility services, along with the conditions for the delivery of these services; and (3) the scope and category of data that is processed in the context of delivering digital mobility services.

The fifth area of regulation pertains to the legal status of the customer as a recipient of digital services within the jurisdiction where the services are to be delivered. The determination of the recipient's status for digital mobility services significantly influences the scope of rights to which the customer is entitled regarding the provision of such services.

3.6 Sources of Law Pertaining to the Provision of Digital Mobility Services as E-commerce Services

3.6.1 E-commerce Regulations that Govern All Legal Relationships

In Section 16(1) of the Digital Services Act (Regulation (EU) 2022/2065, DSA) [82], the European Union legislator has instituted restrictions on hosting service providers mandating them to establish mechanisms that enable any individual or entity to notify them of the presence of specific information on their service that is deemed to constitute illegal content. These mechanisms must be readily accessible and user-friendly, allowing notifications to be submitted exclusively through electronic means. The European Union legislator has delineated in the DSA that hosting service providers occupy a particularly critical role in combatting illegal content on the Internet, as they store information that is provided by and at the request of recipients of the service, thereby rendering it generally accessible to other recipients, often on a substantial scale. The obligation to implement notice-and-action mechanisms should extend, for instance, to file storage and sharing services, website hosting services, advertising servers, and paste-up containers, provided that these services qualify as hosting services encompassed by this regulation (recital 50 of the DSA). As elucidated in the aforementioned Recital 50, all hosting providers are mandated to implement easily accessible and user-friendly notification and action mechanisms that facilitate the notification of specific items of information that the notifying party regards as illegal content. Such mechanisms should be prominently identifiable, located in proximity to the disputed information, and must be at least as easy to locate and utilise as the hosting provider's terms and conditions notification mechanisms.

Polish law incorporates regulations pertaining to e-commerce. Primarily, it is essential to reference the Act of 18 July 2002 [7] on the Provision of Services by Electronic Means. This Act delineates, among other stipulations, the obligation of the service provider to establish the regulations governing the provision of services by electronic means and to make these regulations accessible to the service recipient prior to their agreement to the provision of such services. Article 8(3) of this Act articulates minimum requirements concerning the content of the regulations for the provision of services by electronic means, which include: (1) types and scope of services delivered via electronic means; (2) conditions for the provision of services by electronic means; (3) terms for concluding and terminating agreements related to the provision of services by electronic means; (4) the procedure for lodging complaints. These regulations are obligatory for the provider, who is mandated to deliver services by electronic means in accordance with these stipulations.

In the context of e-commerce regulations relevant to both business-to-consumer and business-to-business relationships, it is also imperative to specify the regulations governing the conclusion of online contracts. Firstly, it is essential to reference Regulation (EU) No 910/2014 (eIDAS) [84], which pertains to electronic identification schemes that have been notified by a Member State, as well as to trust service providers established within the Union (Article 2(1) of eIDAS). The term 'electronic identification scheme' is defined in the Regulation to encompass a system for electronic identification under which electronic identification means are issued to natural or legal persons, or natural persons representing legal persons. The objective of this regulation is to facilitate secure cross-border transactions by establishing a framework for digital identity and digital authentication. This regulation specifically addresses electronic identification (eID) and trust service providers, aiming to eliminate existing barriers that obstruct the seamless use of trust services and the application of electronic identification across all EU Member States. A pivotal aspect of the eIDAS Regulation is the establishment of mutual recognition of electronic identifiers issued by EU countries, contingent on their compliance with specific regulatory criteria and their formal notification to the Commission. This recognition permits secure electronic transactions by certifying that an electronic identifier issued in one Member State is duly valid and recognised in all other Member States. Notably, mutual recognition becomes mandatory for electronic identifiers meeting specified assurance levels, thereby facilitating cross-border interactions and enhancing trust in online services [24].

Issues pertaining to the conclusion of contracts via online platforms are also addressed within the national legislation of the Member States. In the context of Polish law, the Civil Code encompasses the following stipulations:

(1) regulation of the offer process when the offer is submitted online (Article 66(1) of the Polish Civil Code (1964));
(2) stipulation regarding the form and effectiveness of legal acts executed electronically (Article 72 et seq. of the Polish Civil Code, specifically concerning the documentary and electronic form);
(3) provision regarding the utilisation of templates in electronic format (Article 384(4) of the Polish Civil Code).

Furthermore, it is essential to consider the legal obligations relevant to the processing of personal data (Chap. 7), as well as the stipulations regarding the confidentiality of electronic communications (Chap. 7).

3.6.2 E-commerce Regulations that Pertain to the Legal Relationships Between Entrepreneurs and Consumers

Regarding the legal regulations governing the relationship between traders and consumers in the realm of e-commerce, it is essential to first refer to the European Union regulations, specifically the Digital Services Act (DSA) and the General Product Safety Regulation (GPSR) [84]. These provisions are directly enforceable in the Member States. Each of these EU regulations stipulates, among other obligations (refer to Article 11 DSA and Article 22 GPSR), that providers of online marketplaces must appoint a singular point of contact for direct electronic communication with the market surveillance authorities of the Member States concerning product safety matters, particularly for the purpose of order notifications. Furthermore, the EU legislator has mandated that providers of online marketplaces establish a dedicated point of contact to enable consumers to engage directly and expeditiously regarding product safety issues (see Article 12 DSA and Article 22 GPSR). Providers of online marketplaces are obliged to collaborate with market surveillance authorities, traders, and relevant economic operators to facilitate any measures undertaken to eliminate or, if such elimination is unfeasible, to mitigate the risks associated with a product offered online through their services (Article 22(12) GPSR). This provision outlines specific obligations in an open catalogue as provided by the EU legislator.

In reference to Polish law, it is pertinent to highlight a collection of national regulations pertaining to e-commerce. Firstly, attention should be drawn to the Civil Code [106], specifically Articles 385^1 and 385^3, which address the implications of employing unfair terms in contracts. The second provision enumerates an open catalogue of unfair terms. Additionally, the Act of 30 May 2014 [10] on consumer rights warrants mention. This Act delineates, among other matters, the obligations of an entrepreneur when entering into a contract with a consumer; the principles and procedures for concluding a contract with a consumer at a distance and off-premises;

the principles and procedures for exercising the consumer' s right to withdraw from a contract concluded at a distance or off-premises; the principles and procedures for forming a distance contract with a consumer concerning financial services; and the principles and procedures for exercising the rights of a consumer engaged in a contract requiring the transfer of ownership of goods to the consumer, as well as those involved in a contract for the supply of digital content or a digital service. Regarding e-commerce matters, the stipulations of Chapter 3 of this Act, which address the obligations of the entrepreneur in contracts finalised outside the business premises or at a distance, are of significant relevance. In this chapter, the legislator distinguishes legal regulations concerning, inter alia: (1) the entrepreneur' s information obligations in contracts concluded at a distance or beyond the business premises, (2) the information obligations of the online trading platform provider towards the consumer, (3) the model instructions for withdrawal from the contract, (4) the obligation to present information in a legible manner and to supply information on a durable medium, (5) the obligation to provide the consumer with a document confirming the contract concluded outside the business premises or the confirmation of its conclusion, (6) the obligation to indicate information regarding restrictions on delivery and accepted methods of payment for online orders, and (7) the obligation to confirm the conclusion of the distance contract.

In Polish law, consumer rights are safeguarded under the Act of 16 February 2007 [6] on the Protection of Competition and Consumers. It is particularly important to reference Chapter 2 of this Act, which addresses the recognition of standard contracts as unlawful. Within this chapter, Articles 23(b) to 23(d) delineate the authority of the President of the Office of Competition and Consumer Protection to render decisions regarding the unlawfulness of provisions in standard contracts and the associated legal ramifications. Additionally, another legislative measure governing consumer rights is the Act of 23 August 2007 [8] on Counteracting Unfair Market Practices. The term 'market practice' is articulated in Article 2(4) of this Act as any action or omission by an entrepreneur, a particular method of conduct, a statement, or commercial information, especially in the realm of advertising and marketing, that is directly connected to the promotion or acquisition of a product by a consumer. This Act prohibits unfair market practices as stated in Article 3, which are deemed contrary to good customs and significantly distort or could potentially distort the behaviour of the average consumer before, during, or after the formation of a product contract (Article 4(1)). Moreover, the Act establishes a blacklist of unfair market practices that are misleading in all circumstances (Article 7) as well as a blacklist of aggressive market practices (Article 9). It additionally prescribes civil and criminal liability concerning unfair market practices directed at consumers.

The Act of 26 April 2024 [9] concerning the assurance of compliance with the requirements regarding the availability of specific products and services by economic entities was enacted in Poland, with certain provisions having partially come into force. This legislation implements Directive (EU) 2019/882 [30] from the European Parliament and the Council dated 17 April 2019, which pertains to the accessibility requirements for products and services. According to Article 3(2)(point 6), this Act is applicable, among other sectors, to e-commerce services that are offered or provided to consumers. Such services are defined within the Act as being provided or offered at a distance through websites and mobile devices, by electronic means, and at the individual consumer's request for the purpose of concluding a contract.

3.6.3 E-commerce Regulations Applicable to Legal Relationships Between Entrepreneurs

With regard to e-commerce regulations that govern the legal relationships between entrepreneurs, Regulation (EU) 2019/1150 [81] should be emphasised. This regulation has been in effect since 12 July 2020. It pertains specifically to the legal relationships between entrepreneurs, particularly concerning platforms that provide intermediation services and entrepreneurs who utilise these services to sell goods or offer services.

The significance of online intermediation services has been outlined by the EU legislator in recital 2 of the Regulation, which indicates that such services play a crucial role in the provision of services and goods from traders to consumers. Given that online intermediation services are a vital component in facilitating communication with consumers, the EU legislator emphasises the considerable functional and economic dependence, particularly of micro, small, and medium-sised enterprises, on providers of online intermediation services. While Regulation 2019/1150 [81] aims to protect traders using the services of intermediation service providers from unfair practices, it also indirectly safeguards the interests of consumers.

The term 'business user' is defined in Article 2(1) of Regulation (EU) 2019/1150 [81], which states that the purpose for which an entity utilises online intermediation services is to facilitate the making of offers to sell goods or to provide services to consumers. The material scope of this Regulation is outlined in Article 1(2), with exclusions specified in paragraphs (3)–(5).

In Article 2(2) of Regulation (EU) 2019/1150 [81], the EU legislator characterises 'online intermediation services' as services that possess the following cumulative characteristics. Firstly, it qualifies as 'information society services' as defined in point (b) of Article 1(1) of Directive (EU) 2015/1535 [29]. Secondly, these services ought to enable business users to offer goods or services to consumers and facilitate the initiation of direct transactions between those business users and consumers, regardless of where these transactions are ultimately consummated. Finally, these services should be delivered to business users based on contractual relationships between the provider and the user of such services.

In the judgment issued by the Court of Justice in Case C-434/15 on 20 December 2017 [49], the Court of Justice of the European Union (CJEU) determined that the services provided by Uber do not qualify as information society services. In the rationale of its judgment, the Court delineated the subject matter of the service provided by Uber from the perspective of the service recipient. It underscored that when passengers are transported by non-professional drivers using their vehicles, the entity facilitating this intermediation service also offers urban transport services, which are made available primarily through software tools, such as the application pertinent to the main proceedings (see recital 38). Consequently, the Court acknowledged the dual existence of services within Uber's operations, specifically the service of soliciting expressions of interest in transport via an online platform, alongside the transport service itself. The Court concluded that the transport service constituted the dominant service, whereas the other was deemed ancillary. Based on this reasoning, the Court ultimately determined that the services offered by Uber were excluded from the category of information society services.

The territorial scope of application of Regulation 2019/1150 is delineated in Article 1(2). Firstly, traders utilising online intermediation or search services must have their registered office or place of residence within the European Union. Secondly, these traders are required, through the provision of such services, to offer their goods or services to consumers located in the European Union.

The recipients of the obligations outlined in Regulation 2019/1150 [81] are providers of online intermediation services (Article 2(3)). The Regulation specifically applies to entities offering the following categories of online services: online auction services, price comparison websites, online application stores, and online reservation systems. Notably, certain provisions of Regulation 2019/1150 [81], particularly Article 5, also extend to online search engines.

Regulation (EU) 2019/1150 [81] outlines a series of obligations mandated by the European Union legislator for providers of online intermediation services. These obligations include, among others: the establishment of the terms and conditions for providing online intermediation services (Article 3), the limitation or suspension of these services to a specific business user concerning individual goods or services offered by that user (Article 4), and the specification of terms for positioning in online search engines, including the primary parameters that govern positioning and the rationale for prioritising these primary parameters over others (Article 5). Furthermore, in instances where ancillary goods and services, including financial products, are made available to consumers through online intermediation services, either by the provider or by third parties, the provider is required to include in its terms and conditions a comprehensive description of the types of ancillary goods and services offered and to clarify whether, and under what conditions, the business user may also offer its own ancillary goods and services through these intermediation services (Article 6). An additional obligation relates to the inclusion within the terms and conditions of online intermediation service providers of an explicit description of any differentiated treatment that they may extend regarding the goods or services offered to consumers (Article 7). Moreover, regulation (EU) 2019/1150 [81] stipulates the obligation to provide an adequate technical and contractual description, alongside granting traders access to any personal or other data that traders or consumers provide for the use of online intermediation services or that are generated as a result of such services (Article 9). Additionally, Articles 12 and 13 of Regulation (EU) 2019/1150 [81] require the establishment of an internal dispute resolution system between online intermediation platforms and traders, which includes the appointment of independent mediators.

3.7 Regulatory Barriers to the Implementation of Digital Mobility Services

Digital mobility services may be implemented within a jurisdiction provided that there are no regulatory impediments. Such impediments may arise from a total ban on the relevant business activities defined by the provision of specific mobile digital services or from legal constraints that significantly hinder the execution of service activities, rendering the goal of servitisation—aimed at generating value—unclear. A fundamental aim of servitisation is to enhance the competitiveness of products that, by leveraging associated services, create a distinct advantage over market competitors. Consequently, the goal of servitisation is to promote competition within the market of digital mobility services. This goal should also guide legislative efforts in structuring regulations governing the provision of these services. Competition can be maintained by upholding the principle of freedom of establishment, along with its enforcement through specialised public institutions. Any restrictions on the implementation of this freedom should be deemed acceptable only in the interest of significant public concerns, such as state security or public order.

According to [2], transportation network companies, such as Uber, Lyft, and Taxify/Bolt, are leveraging advances in Information and Communication Technology (ICT) and artificial intelligence to provide innovative digital platform mobility solutions, including ride-hailing, ride-sharing, car-sharing, and bike-sharing in urban environments worldwide [101]. These on-demand mobility solutions represent a significant component of the evolving Mobility-as-a-Service (MaaS) ecosystem [50]. The shared mobility market is recognised as one of the fastest-growing segments within the mobility industry. By utilising the advantages of online market platforms, such as Smart digital platform mobility solutions, particularly internet-based ride-hailing, shared mobility (SM) services facilitate the transformation of underutilised societal resources into new business opportunities [3, 22]. Notably, the emergence and success of Uber have catalysed the recognition of the socio-economic potential associated with related services, prompting the introduction of various types of Smart mobility services across numerous countries [22, 105].

However, as stated in the literature [32], the type of service provided by Uber is not unique. In fact, in San Francisco itself, rival companies Lyft and Sidecar already offered app-based services similar to those introduced by Uber in 2010. Lyft continues to be a major competitor in the United States. In South-East Asia, Uber faced competition from companies such as Grab and Ola in India. Grab, headquartered in Singapore as the 'Everyday Everything App', serves as Southeast Asia's equivalent to Uber, DoorDash, Instacart, and Venmo—all on a single platform. In addition to ride-hailing and delivery services, the Grab app provides a wallet feature that enables users to make payments online and in person, send money to family and friends, finance purchases over time, and even purchase travel insurance [91]. Grab operates in Cambodia, Indonesia, Singapore, Malaysia, Myanmar, the Philippines, Thailand, and Vietnam. Ola, a successful Indian ride-hailing company that has revolutionised the cab booking industry, also operates in Australia, New Zealand, and the United Kingdom. In addition to car-sharing and hailing services, Ola provides financial, food, and delivery services [72]. In China, competition has been particularly intense. Uber invested significant resources into this market but was compelled to merge with its chief rival, Didi Chuxing, in 2016 [27]. Despite the scale of its operations, the losses incurred in China and other countries have raised concerns regarding Uber's long-term profitability.

In regard to the global trend of digitalisation, the introduction of Uber, along with its competitors, into the mobility service market signifies the emergence of a new niche: platform-based mobility services [28, 32, 56]. By utilising internet-based mobile technology to connect drivers and users, Uber endeavours to enhance the experience for both parties involved. This development has had a considerable impact on the existing service providers.

These platform-based services reduce the marginal cost associated with providing goods and services while offering more reliable transportation options [62]. Furthermore, the availability of shared mobility services is considerable, as they allow local drivers to utilise their personal vehicles for commercial purposes. Shaheen et al. [96] posited that shared mobility services positively impact society by enhancing job accessibility and fostering social connections. Alonso-Mora et al. [13] demonstrated that these services can help alleviate traffic congestion. In this context, various shared mobility services, including taxi-hailing, ride-hailing, and ridesharing, have been introduced. Taxi-hailing refers to a shared mobility service wherein a customer summons a licensed taxi driver. In the case of ride-hailing, a customer calls upon a driver who is registered to provide such services. Ridesharing, often interchangeably referred to as carpooling, permits customers to share a ride with others. Conversely, accessibility services facilitate participation in activities without necessitating travel or vehicle ownership (e.g., virtual meeting services and e-health services that enable online consultations with healthcare professionals) [86]. In recent years, there has been considerable interest in the potential integration of these services, particularly mobility services, into a user-friendly and seamless offering, conceptualised as Mobility as a Service (MaaS) (see, e.g., [85, 100]).

In South Korea, the introduction of shared mobility services has been obstructed by existing regulatory access barriers that arise from stringent regulations [63]. Both ride-sharing and ride-hailing services are prohibited in South Korea, and global shared mobility services are also banned.

South Korea's ban represented the first formal prohibition of Uber at the national level [70]. This measure primarily arose from regulatory policies [71] and substantial resistance from the taxi industry [45]. Within South Korea, prevailing regulations allow only taxi-hailing services. According to Article 81 of the Passenger Transport Service Act (concerning the Prohibition against Commercial Transport by Non-commercial Motor Vehicles, hereafter referred to as 'PTSA') [58], the provision or rental of motor vehicles for commercial purposes is explicitly prohibited. Moreover, the taxi drivers' union claimed that all ride-sharing services, except for taxi-hailing, violate Article 81 of the PTSA, as they enable unauthorised individuals to provide ride services. The key distinction between taxi-hailing services and other types lies in the requirement that only licensed taxi drivers are permitted to operate, while the latter impose no such qualifications on drivers. Compared to other ride-sharing services, taxi-hailing services faced relatively less opposition from taxi drivers, as they managed to meet regulatory standards while addressing the needs of existing stakeholders. This resulted in the emergence of a taxi-hailing service that exclusively utilises taxi resources [56] and contributed to the formation of a natural monopoly within the industry [45].

As previously noted, the South Korean government has demonstrated a passive approach in loosening regulations within the mobility sector, primarily attributable to significant opposition from entrenched stakeholders, particularly the taxi industry.

3.8 Conclusions

Legal regulation constitutes a vital element influencing the feasibility of transitioning from a product-centric to a service-centric business model. In addition to legal factors, several non-legal aspects also affect this feasibility, including access to adequate resources, necessary organisational changes, and the demand for specific digital mobility services. The law presents risks that can impact the effectiveness of the servitisation process, which can be multi-faceted. First and foremost, it is imperative to indicate the manner in which the principle of economic freedom is upheld in countries where economic activities involving the provision of digital mobility services are to be undertaken. Economic freedom should not be perceived as arbitrariness; rather, it should be associated with the establishment and adherence to legal requirements, which serve as a prerequisite for initiating and conducting economic activities. The establishment of state or state-related monopolies, along with the imposition of legal requirements that are disconnected from market realities, leads to the imposition of legal barriers that restrict the conduct of such economic activities. Legal obligations pertinent to the commencement and execution of economic activities related to the provision of digital mobility services may directly arise from legal norms enshrined in statutory law. Furthermore, the legislator may delegate the responsibility of specifying a legal obligation to the administrative body, thereby granting it the authority to issue discretionary decisions. Alternatively, the legislator may transfer the obligation of clarifying duties arising from legal norms directly to the addressee of these norms. A pertinent example is Regulation (EU) 2016/679, wherein certain obligations are delineated by the addressee of these obligations based on the criterion concerning the violation of the rights or freedoms of natural persons whose personal data is processed. The complexity of the circumstances under which legal obligations emerge poses significant challenges for entrepreneurs contemplating the implementation of digital mobility services.

Regulation can also act as a barrier to experimentation. For instance, it may facilitate the acceleration of the low-emissions transition or address liability concerns related to autonomous driving. Disruptive innovation can raise significant public policy issues, generating a demand for novel regulatory approaches. It is crucial to recognise that establishing a legal framework for the provision of digital mobility services presents a considerable challenge for legislators. The legal framework governing road transport automation, which includes product liability, cybersecurity, and intellectual property, holds substantial significance. The regulatory structure for implementing digital mobility services must also take into account the supervisory responsibilities of regulators. Regarding the use of autonomous vehicles, the existing literature highlights various perspectives. Entrepreneurs act as catalysts for economic transformation by introducing new products, production methods, markets, sources of supply, or industrial combinations, thereby propelling the economy into a state of creative destruction.

It is essential to consider that digital mobility services are provided by suppliers who collaborate with third parties within the established ecosystem. A key feature of the business models linked to the implementation of digital mobility services is the shift of these offerings, as well as their acquisition methods, to the Internet, thereby classifying them as Internet-based services (over-the-top services). A prerequisite for engaging with and purchasing these services is that the customer must have access to the Internet, which constitutes a technical requirement. Nonetheless, both the promotion of digital mobility services and their procurement involving an intermediary entity, specifically a digital platform operator, is subject to the legal regulations governing e-commerce.

References

1. Abduljabbar, R.L., Liyanage, S., Dia, H.: The role of micro mobility in shaping sustainable cities: a systematic literature review. Transp. Res. Part D: Transp. Environ. **92**, 102734 (2021). https://doi.org/10.1016/j.trd.2021.102734
2. Acheampong, R.: Societal impacts of smart, digital platform mobility services—an empirical study and policy implications of passenger safety and security in ride-hailing. Case Stud. Transp. Policy **9**, 302–314 (2021)
3. Acquier, A., Daudigeos, T., Pinkse, J.: Promises and paradoxes of the sharing economy: an organizing framework. Technol. Forecast. Soc. Chang. **125**, 1–10 (2017). https://doi.org/10.1016/j.techfore.2017.07.006
4. Act of 11 September 2015 on insurance and reinsurance activities (consolidated text: Journal of Laws of 2024, item 838, as amended)
5. Act of 14 June 1960 Code of Administrative Procedure (i.e. Journal of Laws 2024, item 572)
6. Act of 16 February 2007 on competition and consumer protection (consolidated text: Journal of Laws of 2024, item 1616)
7. Act of 18 July 2002 on the provision of services by electronic means, consolidated text: Journal of Laws of 2024, item 1513, as amended
8. Act of 23 August 2007 on counteracting unfair market practices (consolidated text: Journal of Laws of 2023, item 845)
9. Act of 26 April 2024 on ensuring compliance with the requirements for the availability of certain products and services by economic entities (Journal of Laws, item 731)
10. Act of 30 May 2014 on consumer rights (consolidated text: Journal of Laws of 2024, item 1796)
11. Act of 8 March 1990 on local government (consolidated text: Journal of Laws of 2024, item 1465, as amended)

12. Alexy, R.: A Theory of Constitutional Rights (J. Rivers, trans.). Oxford University Press, Oxford, UK (2002) (Original Work: Theorie der Grundrechte published in Baden-Baden, 1985)
13. Alonso-Mora, J., Samaranayake, S., Wallar, A., Frazzoli, E., Rus, D.: On-demand high-capacity ride-sharing via dynamic trip-vehicle assignment. Proc. Natl. Acad. Sci. U. S. A. **114**, 462–467 (2017). https://doi.org/10.1073/pnas.1611675114
14. Antikainen, R., Alhola, K., Jaaskelainen, T.: Experiments as a means towards sustainable societies—lessons learnt and future outlooks from a Finnish perspective. J. Clean. Prod. **169**(15), 216–224 (2017). https://doi.org/10.1016/j.jclepro.2017.06.184
15. Austin, J.: In: Campbell, R. (ed.) Lectures on Jurisprudence, or the Philosophy of Positive Law, 4th edn., vol. 1, rev. John Murray, UK, London (1873)
16. Bastl, M., Johnson, M., Lightfoot, H., Evans, S.: Buyer-supplier relationships in a servitized environment: an examination with Cannon and Perreault's framework. Int. J. Oper. Prod. Manag.Oper. Prod. Manag. **32**(6), 650–675 (2012)
17. Benedettini, O., Neely, A., Swink, M.: Why do servitized firms fail? A risk-based explanation. Int. J. Oper. Prod. Manag.Oper. Prod. Manag. **35**(6), 946–979 (2015)
18. Bielecki, L.: Zasada wolności gospodarczej w okresie transformacji ustrojowej w Polsce, [w:] Studia Iuridica Lublinensia, t. 1, p. 63. Lublin (2003)
19. Biernat, S., Wasilewski, A.: Wolność gospodarcza w Europie, pp. 212–214. Kraków (2000)
20. Bjørnskov, C., Foss, N.J.: Economic freedom and entrepreneurial activity: some cross-country evidence. Public Choice **134**, 307–328 (2008). https://doi.org/10.1007/s11127-007-9229-y
21. Boons, F., Carlos, M., Jaco, Q., Marcus, W.: Sustainable innovation, business models and economic performance: an overview. J. Clean. Prod. **45**, 1–8 (2013). https://doi.org/10.1016/j.jclepro.2012.08.013
22. Casprini, E., Di Minin, A., Paraboschi, A.: How do companies organize nascent markets? The BlaBlaCar case in the inter-city shared mobility market. Technol. Forecast. Soc. Chang. **144**, 270–281 (2019). https://doi.org/10.1016/j.techfore.2018.01.012
23. Christensen, C.M., Grossman, J., Hwang, J.: Innovator's Prescription: A Disruptive Solution for Health Care. McGraw Hill, New York (2009)
24. Commission 2024. https://digital-strategy.ec.europa.eu/pl/policies/eidas-regulation. Accessed 1 Apr 2025
25. Consolidated text: Council Regulation (EC) No 2157/2001 of 8 October 2001 on the Statute for a European company (SE), ELI: http://data.europa.eu/eli/reg/2001/2157/2013-07-01. Accessed 21 Feb 2025
26. Constitution of the Republic of Poland of 2 April 1997 (Journal of Laws No. 78, item 483, as amended)
27. Crabtree, J.: Didi Chauxing took on Uber and won. Now it's taking on the world, Wired 9.02.20218, Didi Chuxing took on Uber and won. Now it's taking on the world | WIRED. Accessed 22 Feb 2025
28. Cramer, J., Krueger, A.B.: Disruptive change in the taxi business: the case of Uber. Am. Econ. Rev. **106**, 177–182 (2016). https://doi.org/10.1257/aer.p20161002
29. Directive (EU) 2015/1535 of the European Parliament and of the Council of 9 September 2015 laying down a procedure for the provision of information in the field of technical regulations and of rules on Information Society services (OJ L 241, 17.9.2015, p. 1)
30. Directive (EU) 2019/882 of the European Parliament and of the Council of 17 April 2019 on the accessibility requirements for products and services (Text with EEA relevance), PE/81/2018/REV/1, OJ L 151, 7 Jun 2019, pp. 70–115
31. Dorociak, M., Grzeszczak, R., Klemt, T., Szmigielski, A., Szpyrka, M., Wajda, P.: Grzeszczak, R. (ed.) Wolność gospodarcza a regulacja rynków na przykładzie wpływu unijnych i krajowych regulatorów rynków na działalność przedsiębiorców, pp. 14–22. Warszawa (2020)
32. Dudley, G., Banister, D., Schwanen, T.: The rise of Uber and regulating the disruptive innovator. Polit. Q. **88**, 492–499 (2017). https://doi.org/10.1111/1467-923X.12373
33. European Convention on Human Rights (ECHR). https://www.eur-lex.europa.eu/legal-content/EN/TXT/?uri=LEGISSUM:eu_human_rights_convention. Accessed 21 Feb 2025

34. Freeman, C., Perez, C.: Structural crisis of adjustment, business cycles and investment behaviour. In: Dosi, G., Freeman, C., Nelson, R., Silverberg, G., Soete, L. (eds.) Technical Change and Economic Theory, pp. 38–66. Pinter, London (1988)
35. Geels, F.W.: Technological transitions as evolutionary reconfiguration processes: a multi-level perspective and a case-study. Res. Policy **31**(8–9), 1257–1274 (2002). https://doi.org/10.1016/S0048-7333(02)00062-8
36. Geels, F.W., Kemp, R.: Dynamics in socio-technical systems: typology of change processes and contrasting case studies. Technol. Soc. **29**(4), 441–455 (2007). https://doi.org/10.1016/j.techsoc.2007.08.009
37. Geiger, T.: Vorstudien zu einer Soziologie des Rechts [Preliminary Studies to Sociology of Law], 2nd edn. Duncker & Humblot, Germany, Berlin (1970)
38. Georghiou, L., Metcalfe, J.S., Gibbons, M., Ray, T., Evans, J.: Post-Innovation Performance, Technological Development and Competition. MacMillan, London (1986)
39. Hailemariam, A., Ivanovski, K.: Greener horizons: economic freedom as a catalyst for the global energy transition. Energy Res. Soc. Sci. **124**, 104058 (2025). https://doi.org/10.1016/j.erss.2025.104058
40. Holzer, B., Millo, Y.: From risks to second-order dangers in financial markets: unintended consequences of risk management systems. New Polit. Econ. **10**(2), 223–245 (2005)
41. Horubski, K.: Wolność prowadzenia działalności gospodarczej w świetle art. 16 Karty praw podstawowych—Wybrane problemy. In: Kiczka, K., Kocowski, T., Małecki W. (eds.) Przegląd Prawa i Administracji, t. CXIV, p. 499. Wrocław (2018)
42. Huntingford, C., Jeffers, E.S., Bonsall, M.B., Christensen, H.M., Lees, T.: Machine learning and artificial intelligence to aid climate change research and preparedness (2019); Spring, M., Araujo, L.: Product biographies in servitisation and the circular economy. Ind. Mark. Manag. 60, 126–137 (2017)
43. Jagielski, J., Piecha, J.: In: Jagielski, J., Wierzbowski, M. (eds.) Prawo administracyjne, Warszawa, p. 175. Wolters Kluwer (2020)
44. Jakimowicz, W.: Bezpośrednie stosowanie przepisów konstytucji przez organy administracji publicznej. Część druga: Stosowanie norm-reguł i stosowanie norm-zasad, PPP 2008/6, p. 17
45. Jin, D.W.: Car-sharing business and passenger transport service act art. 81: based on the perspective on sharing economy and consumer protection. Hongik Law Rev. **17**, 1–26 (2016)
46. Johnstone, P., Rogge, K.S., Kivimaa, P., Fratini, Ch.F., Primmer, E., Stirling, A.: Waves of disruption in clean energy transitions: Sociotechnical dimensions of system disruption in Germany and the United Kingdom. Energy Res. Soc. Sci. **59**, 101287 (2020). https://doi.org/10.1016/j.erss.2019.101287
47. Judgment of the Constitutional Court of 19 January 2010, SK 35/08, OTK-A 2010/1/2
48. Judgment of the Constitutional Court of 5 June 2014, K 35/11, OTK-A 2014/6/61
49. Judgment of the Court of Justice in Case C-434/15 of 20 December 2017, Asociación Profesional Elite Taxi v.Uber Systems Spain SL, C-434/15, ECLI:EU:C:2017:981
50. Kamargianni, M., Matyas, M.: The business ecosystem of mobility-as-a-service. In: Transportation Research Board, vol. 96. Transportation Research Board (2017). https://discovery.ucl.ac.uk/id/eprint/10037890/1/a2135d_445259f704474f0f8116ccb625bdf7f8.pdf
51. Kelsen, H.: Reine Rechtslehre [Pure Theory of Law], 2nd edn. Deuticke, Austria, Vienna (1960)
52. Kelsen, H.: Reine Rechtslehre: Einleitung in die rechtswissenschaftliche Problematik [Pure Theory of Law: Introduction into Jurisprudential Difficulty. Deuticke, Germany, Leipzig & Austria, Vienna (1934)
53. Kemp, R.: Technology and the transition to environmental sustainability. The problem of technological regime shifts. Futures **26**(10), 1023–1046 (1994)
54. Kemp, R., Rip, A., Schot, J.W.: Constructing transition paths through the management of niches. In: Garud, R., Karnoe, P. (eds.) Path Dependence and Creation, pp. 269–299. Lawrence Erlbaum, Mahwah, NJ (2001)
55. Khanra, S., Dhir, A., Parida, V., Kohtamaki, M.: Servitization research: a review and bibliometric analysis of past achievements and future promises. J. Bus. Res. **131**, 151–166 (2021). https://doi.org/10.1016/j.jbusres.2021.03.056

56. Kim, K., Baek, C., Lee, J.D.: Creative destruction of the sharing economy in action: the case of Uber. Transport. Res. Policy Pract. **110**, 118–127 (2018). https://doi.org/10.1016/j.tra.2018.01.014
57. Kim, K., Kim, T., Cho, S., Jang, H.: Deregulation Strategies for Promoting Innovation of Technology and Services in the Transport Sector in the Era of the 4th Industrial Revolution. The Korea Transport Institute (2019). https://elaw.klri.re.kr/eng_mobile/ganadaDetail.do?hseq=42804&type=abc&key=PASSENGER%20TRANSPORT%20SERVICE%20ACT&param=P. Accessed 12 Feb 2025
58. Klecha, K.: Wolność działalności gospodarczej w Konstytucji RP, p. 174. Warszawa (2009)
59. Kosikowski, C.: Wolność działalności gospodarczej i jej ograniczenia w praktyce stosowania Konstytucji RP. In: Kosikowski, C. (ed.) Zasada ustroju społecznego i gospodarczego w procesie stosowania Konstytucji, pp. 43–45. Warszawa (2005)
60. Kosiński, E.: Aspekt prawny wolności gospodarczej. In: Kwartalnik Prawa Publicznego 2003, nr 4, p. 31. Warszawa (2003)
61. Kulesza, C.: Zasada wolności działalności gospodarczej w orzecznictwie Trybunału Konstytucyjnego, Temmida 2, Białystok, p. 155 (2006)
62. Le Vine, S., Polak, J.: Introduction to special issue: new directions in shared-mobility research. Transportation **42**, 407–411 (2015). https://doi.org/10.1007/s11116-015-9603-4
63. Lee, J., Park, G., Kim, K.: Transition failure strengthening regime stability in socio-technical systems: a case study of shared mobility market in South Korea. Telematics Inform. **70**, 101814 (2022). https://doi.org/10.1016/j.tele.2022.101814
64. Liu, H.Y.: Irresponsibilities, inequalities and injustice for autonomous vehicles. Ethics Inf. Technol. **19**(3), 193–207 (2017)
65. Long, T.B., Arnold, L., Vincent, B.: Critical success factors for the transition to business models for sustainability in the food and beverage industry in the Netherlands. J. Clean. Prod. **175**, 82–95 (2018). https://doi.org/10.1016/j.jclepro.2017.11.067
66. Rossi, L.S.: How fundamental are fundamental principles? Primacy of the EU Law, principles of the national constitutions and fundamental rights after Lisbon. In: Venturini, G., Bariatti, S. (eds.) Diritti individuali e giustizia internazionale: Liber Fausto Pocar 801–22 (2009)
67. Luhmann, N.: Rechtssoziologie [Sociology of law], 4th edn. VS Verlag fuer Sozialwissenschaften, Germany, Wiesbaden (2008)
68. Malerba, F., Orsenigo, L.: Technological regimes and sectoral patterns of innovative activities. Ind. Corp. Change **6**(1), 83–117 (1997)
69. Mathieu, V.: Service strategies within the manufacturing sector: benefits, costs and partnership. Int. J. Serv. Ind. Manag. **12**(5), 451–475 (2001)
70. McSpadden, K.: Setback for Uber as South Korea bans private taxis. Times (2015). Retrieved from: http://time.com/3901066/uber-ban-south-korea-taxi-uberblack-uberx
71. Moon, M.J.: Government-driven sharing economy: lessons from the sharing city initiative of the Seoul metropolitan government. J. Dev. Soc. **33**, 223–243 (2017). https://doi.org/10.1177/0169796X17710076
72. Nandi, S.: Ola Case Study—The Success Story of Ola! Business Rule, 17 Aug 2023. https://thebusinessrule.com/ola-case-study-the-success-story-of-ola/. Accessed 12 Feb 2025
73. Nelson, R.R., Winter, S.G.: In search of useful theory of innovation. Res. Policy **6**, 36–76 (1977)
74. Nelson, R.R., Winter, S.G.: An Evolutionary Theory of Economic Change. Belknap Press, Cambridge, MA (1982)
75. Nikitas, A., Njoya, E.T., Dani, S.: Examining the myths of connected and autonomous vehicles: analysing the pathway to a driverless mobility paradigm. Int. J. Autom. Technol. Manag. **19**(1–2), 10–30 (2019)
76. Oermann, M., Ziebarth, L.: Interpreting code—adapting the methodology to analyse the normative contents of law for the analysis of technology. Comput. Law Secur. Rev.. Law Secur. Rev. **31**(2), 257–267 (2015). https://doi.org/10.1016/j.clsr.2015.01.008
77. Olszanowski, J.: 3.2.1.2. Treść zasady. In: Łaszczyca, G., Matan, A., Piątek, W., Tarno, J.P. (eds.) System Prawa Administracyjnego Procesowego. Tom II. Część 2. Zasady ogólne postępowania administracyjnego. Warszawa (2018)

78. Prakken, H.: On the problem of making autonomous vehicles conform to traffic law. Artif. Intell. Law. Intell. Law **25**(3), 341–363 (2017)
79. Przybysz, P.M.: [w:] Kodeks postępowania administracyjnego. Komentarz aktualizowany, LEX/el. 2024, art. 6
80. Regulation (EU) 2016/679 of the European Parliament and of the Council of 27 April 2016 on the protection of natural persons with regard to the processing of personal data and on the free movement of such data, and repealing Directive 95/46/EC, OJ L 119, 4 May 2016, pp. 1–88
81. Regulation (EU) 2019/1150 of the European Parliament and of the Council of 20 June 2019 on promoting fairness and transparency for business users of online intermediation services (Text with EEA relevance), PE/56/2019/REV/1, OJ L 186, 11 Jul 2019, pp. 57–79
82. Regulation (EU) 2022/2065 of the European Parliament and of the Council of 19 October 2022 on a Single Market For Digital Services and amending Directive 2000/31/EC (Digital Services Act) (Text with EEA relevance), PE/30/2022/REV/1, OJ L 277, 27 Oct 2022, pp. 1–102
83. Regulation (EU) 2023/988 of the European Parliament and of the Council of 10 May 2023 on general product safety, amending Regulation (EU) No 1025/2012 of the European Parliament and of the Council and Directive (EU) 2020/1828 of the European Parliament and the Council, and repealing Directive 2001/95/EC of the European Parliament and of the Council and Council Directive 87/357/EEC (Text with EEA relevance) PE/79/2022/REV/1, OJ L 135, 23 May 2023, pp. 1–51
84. Regulation (EU) No 910/2014 of the European Parliament and of the Council of 23 July 2014 on electronic identification and trust services for electronic transactions in the internal market and repealing Directive 1999/93/EC, OJ L 257, 28 Aug 2014, pp. 73–114
85. Ringenson, T., Kramers, A.: Mobility as a service and the avoid-shift-improve approach. In: Advances and New Trends in Environmental Informatics. ENVIROINFO 2021, pp. 217–234 (2021). https://doi.org/10.1007/978-3-030-88063-7_14
86. Ringenson, T., Arnfalk, P., Kramers, A., Sopjani, L.: Indicators for promising accessibility and mobility services. Sustainability (Switzerland) **10**(8), 1–19 (2018). https://doi.org/10.3390/su10082836
87. Rip, A., Kemp, R.: Towards a Theory of Socio-Technical Change, mimeo UT, report prepared for Batelle Pacific North West Laboratories, Washington, DC (1996)
88. Rip, A., Kemp, R.: Technological change. In: Rayner, S., Malone, L. (eds.) Human Choice and Climate Change, vol. 2. Resources and Technology, pp. 327–400. Batelle Press, Washington, DC (1998)
89. Rosenberg, N.: Path-dependent aspects of technological change. In: Exploring the Black Box. Technology, Economics, and History, pp. 9–23. Cambridge University Press, Cambridge (1994)
90. Rosenbloom, D.: A clash of socio-technical systems: exploring actor interactions around electrification and electricity trade in unfolding low-carbon pathways for Ontario. Energy Res. Soc. Sci. **49**, 219–232 (2019). https://doi.org/10.1016/j.erss.2018.10.015
91. Salvucci, J.: What is Grab? Southeast Asia's post-Uber "everything app". TheStreet, 9 Apr 2024. https://finance.yahoo.com/news/meet-grab-southeast-asia-post-011400820.html?guccounter=1&guce_referrer=aHR0cHM6Ly93d3cuYmluZy5jb20v&guce_referrer_sig=AQAAANy9Q00sbss0ZMLWwcNfqEadgO8lVwMFK8w6_LPuA4VNYIV5ih9GyLEHh0VLkbdfTv8uujTWO6VFRuDdVSNZkSgp7n79iSglvt5V8XPyIR74IWbCtwqN2WH0jyIDXTm2UE3qN8lrc_G3XtgmeNstCJye9aaip0Qpkgi2gtAfikWk. Accessed 12 Feb 2025
92. Sanitt, A., Evans, M., Saddar, S., Evans, H., Hidaka, S.: Autonomous Vehicles: The Legal Landscape of DSRC in the United Kingdom in Autonomous Vehicles: The Legal Landscape of Dedicated Short Range Communication in the US, UK and Germany. Report, Norton Rose Fulbright (2017)
93. Schot, J., Hoogma, R., Elzen, B.: Strategies for shifting technological systems. The case of the automobile system. Futures **26**, 1060–1076 (1994)
94. Schumpeter, J.A.: Capitalism, Socialism, and Democracy. Harper Perennial, London (1942)

95. Segerstrom, P.S., Anant, T.C.A., Dinopoulos, E.: A Schumpetenan model of the product life cycle. Am. Econ. Rev. **80**, 1077–1091 (1990)
96. Shaheen, S.A.: Mobility and the sharing economy. Transp. Policy **51**, 141–142 (2016). https://doi.org/10.1016/j.tranpol.2016.01.008
97. Śledzińska-Simon, A.: Analiza proporcjonalności ograniczeń konstytucyjnych praw i wolności, p. 16. Wrocław (2019)
98. Sługocka, M.: Wolność działalności gospodarczej jako norma- zasada, wyd: Prace prawnicze, p. 136. Warszawa (2012)
99. Smith, A.: Translating sustainability between green niches and socio-technical regimes. Technol. Anal. Strateg. Manag. **19**, 427–450 (2007)
100. Smith, G., Hensher, D.A.: Towards a framework for mobility-as-a-service policies. Transp. Policy **89**, 54–65 (2020). https://doi.org/10.1016/j.tranpol.2020.02.004
101. Stehlin, J., Hodson, M., McMeekin, A.: Platform mobilities and the production of urban space: toward a typology of platformization trajectories. Environ. Plan. A: Econ. Space 0308518X19896801 (2020)
102. Sypniewski, D., Legierska, U., Małkowska, A.: Ograniczenia wolności podejmowania działalności gospodarczej, red. D. Sypniewski, Monografia Prawnicza, p. 23. Warszawa (2016)
103. Sypniewski, D.: Ograniczenia wolności podejmowania działalności gospodarczej, p. 24. Warszawa (2016)
104. Szafrański, A.: Konstytucyjna wolność gospodarcza na tle historii idei i gospodarki, BECK, Warszawa 2018, pp. 123.
105. Tauscher, K., Laudien, S.M.: Understanding platform business models: a mixed methods study of marketplaces. Eur. Manag. J.Manag. J. **36**(3), 319–329 (2018). https://doi.org/10.1016/j.emj.2017.06.005
106. The Act of 23 April 1964—the Civil Code (consolidated text: Journal of Laws of 2024, item 1061, as amended)
107. The Charter of Fundamental Rights of the European Union, 2012/C 326/02, OJ C 326, 26 Oct 2012, pp. 391–407
108. Treaty on European Union (OJ C 191, 29.7.1992, pp. 1–112)
109. Treaty on the Functioning of the European Union, European Union, Consolidated versions of the Treaty on European Union and the Treaty on the functioning of the European Union—Charter of Fundamental Rights of the European Union, Publications Office (2012)
110. Usai, A.: The freedom to conduct a business in the EU, its limitations and its role in the European legal order: a new engine for deeper and stronger economic, social, and political integration. German Law J. **14**(09), 1868–1888 (2019)
111. Van de Poel, I.: The transformation of technological regimes. Res. Policy **32**, 49–68 (2003). PII: S0048-7333(01)00195-0
112. Van den Ende, J., Kemp, R.: Technological transformations in history: how the computer regime grew out of existing computing regimes. Res. Policy **28**, 833–851 (1999)
113. von Ungern-Sternberg, A.: Autonomous driving: regulatory challenges raised by artificial decision-making and tragic choices. In: Barfield, W., Pagallo, U. (eds.) Research Handbook on the Law of Artificial Intelligence. Edward Elgar (2018)
114. Wautelet, Y., Manuel, K., Samedi, H., Stephan, P.: Developing a multi-agent platform supporting patient hospital stays following a socio-technical approach: management and governance benefits. Telemat. Inform. **35**(4), 854–882 (2018). https://doi.org/10.1016/j.tele.2017.12.013
115. Windahl, C., Lakemond, N.: Integrating solutions from a service-centred perspective: applicability and limitations in the capital goods industry. Ind. Mark. Manag.Manag. **39**(8), 1278–1290 (2010)
116. Wiseman, Y.: Driverless cars will make Union Stations obsolete. Open Transp. J. **13**, 109–115 (2019). https://doi.org/10.2174/1874447801913010109
117. Won Bogdandy, A.: Founding principles. In: von Bogdandy, A., Bast, J. (eds.) Principles of European Constitutional Law, 2nd edn., p. 22. Hart Publishing (2009)

118. Zaradkiewicz, K., Safjan, M., Bosek, L.: Konstytucja, t. 1., BECK, p. 554. Warszawa (2016)
119. Zhang, W., Banerji, S.: Challenges of servitization: a systematic literature review. Ind. Mark. Manag. Manag. **65**, 217–227 (2017). https://doi.org/10.1016/j.indmarman.2017.06.003
120. Zhang, X., You, W., Shan, L.: Exploring short-form video application addiction: socio-technical and attachment perspectives. Telemat. Inform. **42**, 101243 (2019). https://doi.org/10.1016/j.tele.2019.101243
121. Zimmermann, J.: Prawo administracyjne, p. 54. Warszawa (2019)
122. Zippelius R von.: Juristische Methodenlehre [Juridical Doctrine Methode], 10th edn. C.H. Beck, Germany, Munich (2006)
123. Żurawik, A.: „Interes publiczny", „interes społeczny" i „interes społecznie uzasadniony". Próba dookreślenia pojęć, wyd: Ruch prawniczy, ekonomiczny i socjologiczny, Poznań, pp. 57–59 (2014)
124. Fang, E., Palmatier, R.W., Steenkamp, J.B.: Effect of service transition strategies on firm value. J. Mark. **72**(September), 1–14. (2008)
125. Pawar, K. S., Beltagui, A., Riedel, J. C. K. H.: The PSO triangle: Designing product, service and organization to create value. Internat. J. Oper. Produ. Manag. **29**(5), 468–493 (2009) http://dx.doi.org/10.201108/01443570910953595
126. Beuren, F. H., Ferreira, M. G. G., Miguel, P. A. C.: Product-service systems: A literature review on integrated products and services. J. Clean. Prod. **47**, 222–231 (2013)
127. van Waes, A., Jacco, F., Koen, F., Jeroen, P.J., Rob, R.: Business model innovation and socio-technical transitions. A new prospective framework with an application to bike sharing. J. Clean. Prod. 195, 1300–1312 (2018) https://doi.org/10.1016/j.jclepro.2018.05.223
128. Dosi, G., Malerba, F., Orsenigo, L.: Evolutionary regimes and industrial dynamics. In: Magnusson, L. Ed. Evolution ary and Neo-Schumpeterian Approaches to Economics. Kluwer, Boston (1993)
129. Sypniewski, D.: Ograniczenia wolności podejmowania działalności gospodarczej, Warszawa (2016)

Chapter 4
Legal Aspects of Digital Mobility Services Provided Within an Ecosystem

4.1 Provision of Digital Mobility Services Within an Ecosystem

The provision of digital mobility services necessitates collaborative interaction between vehicle manufacturers and third-party entities. Turienzo et al. [16] underscore the significance of establishing alliances with technological and data management firms, as well as personalising services, which serve as fundamental pillars of such offerings. Furthermore, entrepreneurs focused on insurance, maintenance, and vehicle care may be required to shift from Business-to-Consumer (B2C) models to Business-to-Business (B2B) environments. This transition will compel them to engage in negotiations with corporate entities rather than end customers, thereby altering the traditional dynamics of cooperation.

These entrepreneurs operate within a defined ecosystem. Jacobides et al. [16] provide a definition of this concept, which encompasses 'interacting organisations' facilitated by modularity. These organisations are not organised hierarchically and are interconnected due to the fact that their joint investments cannot be easily redirected to alternative ventures. Ecosystems generate value by empowering managers to navigate their interdependencies through roles that correspond with shared principles, thereby obviating the necessity for customised contracts with each partner.

This definition considers the pertinent actor structure [16, 16], which includes actors, governance, linkages, roles, and activities [16]. According to Lampón et al. [16], interdependence is crucial to the reciprocal influence of participants on their capacity to generate value. In this context, firms endeavour to align this interdependence structure among ecosystem participants to formulate a central value proposition.

The ecosystem encompasses the function of governance, which must be broadened to include the responsibilities of legislators in facilitating the provision of digital mobility services, as well as those of regulators in the enforcement of the law. The academic literature recognises that the mobility industry, along with its regulatory

A. Krasuski, *Legal Challenges of a Transformation from a Product-Centric to a Service-Centric Business Model and Procedures*, Studies in Systems, Decision and Control 610, https://doi.org/10.1007/978-3-031-98901-8_4

bodies, is at a crucial juncture concerning the implementation of these innovative mobility business models, which promote significant advancements in energy efficiency and low-emission urban accessibility while concurrently generating business value [7, 16, 16]. Reduced entry barriers, increased competition, and the growing digital culture among consumers constitute the most considerable threats [16].

The collaboration among manufacturers of semi- and fully autonomous vehicles and third-party entities within the ecosystem culminates in the formulation of diverse business models, notably illustrated by the 'service-centric' approach. Section 4.2 provides an in-depth examination of selected digital mobility business models. In Sect. 4.3, a comprehensive definition of digital mobility services is presented, accompanied by an analysis of the various categories of services. Given the absence of specific regulations governing this category of services, Sect. 4.4 explores the legal implications associated with the different types of digital mobility services.

4.2 Disruptive Servitisation in Mobility Involving Digital Business Models and Their Ecosystems Pertaining to Autonomous Vehicles

The development of business models within an ecosystem rests on foundational pillars that undergo rigorous scientific inquiry. According to Kohtämaki et al. [16], one of the fundamental pillars of the servitisation process, which embodies a service-centric approach, is digitalisation. This element plays a significant role in the evolution of these models towards the achievement of 'autonomous solutions'. Other crucial pillars encompass the customisation of offerings and the pricing of outcomes. This transformative process, known as servitisation, necessitates the creation of technology-enabled business models that facilitate customer knowledge-based services throughout the entire life cycle of manufactured products [14].

Leminen et al. [16] enumerate four digital servitisation business models within the framework of the characteristics of autonomous vehicle solutions. These business models were established based on specified criteria. Initially, the criterion of 'autonomy' distinguishes between 'semi-autonomous' and 'fully autonomous' vehicles. The subsequent criterion, 'complexity', pertains to the classification into 'single vehicle' and 'fleet'. Vehicles are integrated into a fleet (or a swarm) utilising digital technologies and data. Such a fleet can perform tasks that exceed the capabilities of an individual vehicle.

Digital mobility services can be offered through the following business models as enumerated by Leminen et al. [16]:

(1) 'safety-as-a-service' business model;
(2) 'efficiency-as-a-service' business model;
(3) 'capacity-as-a-service' business model;
(4) 'flexibility-as-a-service' business model.

The initially proposed digital servitisation business model focuses on providing driver functionalities to enhance road safety. The importance of road safety in the context of autonomous vehicles has been highlighted in the literature [16, 16]. This model applies to those levels of automation in motor vehicles that involve human participation in performing the dynamic driving task. Specifically, this encompasses vehicles classified as levels 1 to 3 of automation, as outlined by SAE International [16]. However, if the vehicles are to operate autonomously, the range of automation levels is limited to the third level. For automation levels 4 and 5, an autonomous vehicle can undertake the entire dynamic driving task, whereas at level 5, this capability includes all types of roads.

The 'safety-as-a-service' business model involves assisting the driver in operating the vehicle during semi-autonomous driving modes (1–3) by offering functionalities that aid navigation within the designated operational driving domain, which includes driving straight ahead, braking, and executing turns. As noted by Leminen et al. [16], the realm of digital mobility services generates customer value by recognising that a driving support system reduces vehicle-on-vehicle collisions through automatic braking, thereby preventing or lessening the impact of rear-end collisions, while providing a comprehensive outlook on future travel, characterised as autonomous, electric, connected, and safe.

The enhancement of safety on the driver's side must be assessed within the broader context of overall road safety, which includes the well-being of other road users and pedestrians. The adoption of digital mobility services by drivers positively contributes to the safety of surrounding road users. Implementing this specific business model has the potential to ease traffic congestion and promote smoother vehicular flow.

The second proposed servitisation business model, referred to as 'efficiency-as-a-service', focuses on semi-autonomous vehicles, such as trucks, that operate within a platooning system. This model specifically targets the driver of the lead vehicle in the platoon, while the other trucks operate autonomously without any driver input. These vehicles automatically adjust their driving parameters—such as direction, speed, and following distance—based on the behaviour of the lead vehicle. They communicate utilising various connectivity tools. The lead driver assumes a multifaceted role, requiring meticulous navigation of the vehicle to ensure that other members of the platoon can adapt to any manoeuvres without jeopardising their safety or that of other road users. In semi-autonomous vehicles (Level 3 automation), the transfer of steering control to the automated driving system (ADS) is temporary; thus, a human driver must supervise the steering process and may need to reclaim control upon the request of the ADS. For customers, the value of this model resides in its capacity to transport a greater volume of cargo to the same destination both safely and efficiently, thereby minimising fuel consumption and the need for additional drivers. To effectively implement this trucking method, real-time data from connected fleet trucks must be processed, thereby enhancing lane capacity and enabling collision-free cargo transport [16].

The third proposed business model is known as 'capacity-as-a-service'. This model pertains to the utilisation of level 4 (high automation) and level 5 (full automation) vehicles. Within these levels of automation, human operators are not tasked with driving responsibilities. This business model presupposes the deployment of vehicles exhibiting high and full automation within confined environments, such as mines, university campuses, and airports. Consequently, the model presented by Leminen et al. [16] assumes the application of autonomous level 5 vehicles in a more limited context than what is implied by the capabilities associated with this level of automation, which encompasses the ability to navigate in all road conditions.

To facilitate the movement along designated routes within confined work environments, it is imperative that vehicles be equipped with, among other components, global navigation satellite positioning systems, advanced sensors, and integrated controllers that are operated remotely from a central control room. This setup is specifically designed for operation by a single operator in the context of unmanned hauling. Such functionalities enhance the utility functions of vehicles, particularly their capacity. Furthermore, the implementation of these functionalities significantly impacts cost efficiency, safety, security, and precision. Remotely operated trucks mitigate risks, reduce costs, minimise emissions, decrease tire wear, and lower fuel consumption, thereby enhancing productivity, especially through improved utilisation of existing capacity. Autonomous delineated solutions eliminate the safety risks associated with erratic behaviour that contributes to numerous mining accidents.

Leminen et al. [16] utilise the term 'autonomously demarcated solutions' to characterise this category of vehicle. These refer to driverless vehicles, in which the driver is not accountable for intervening in the driving process, instead, autonomously demarcated solutions may be operated externally. In the context of utilising Autonomous Vehicles (AVSs) as a capacity-as-a-service solution, enhanced utilisation of capacity results from increased operational hours per truck, reduced wear and tear, leading to an extended lifespan of equipment and vehicles, as well as more efficient utilisation of personnel resources. Capacity-as-a-service is evolving to incorporate versatile value-adding service elements that extend beyond mere 'outdoor operations'. The underlying business model is predicated upon a transition from mere monitoring to employing AVSs, which involves the integration of value-adding service elements, such as predictive maintenance and comprehensive factory optimisation, in pursuit of maximising output attainable with a fixed level of personnel, installations, and equipment. Enhanced capacity utilisation is achieved through extended operational hours per vehicle, reduced wear and tear, resulting in a prolonged lifespan of equipment and vehicles, alongside more efficient and effective utilisation of personnel resources.

The fourth business model pertains to autonomous swarmed solutions and assumes the 'flexibility' of utilising autonomous vehicles as a service, referred to as the 'flexibility-as-a-service' business model. This particular model is specifically designated for autonomous vehicles, encompassing the fourth and fifth levels of automation, which signify full automation. When considering the fifth level of automation, it is stipulated that autonomous vehicles can be employed without the geographical limitations previously established in the earlier business model. The third characteristic, aside from the autonomous operation of level 5 vehicles across all road conditions, is the applicability of this model to swarm solutions. As noted by Leminen et al. [16], vehicles collaborate to determine routes, construct structures, and assign tasks, as well as adapt to changes in the environment, including obstacles present on open routes. In semi-autonomous platooning solutions, vehicles are organised into a platoon, which is governed and managed semi-autonomously by a driver. Conversely, autonomous swarmed solutions operate fully automated fleets without any involvement from a driver or external operator. Within autonomous swarm solutions, vehicles function collectively and respond to prospective needs as a cohesive unit or swarm by restructuring their operational procedures.

According to Leminen et al. [16], such vehicles share data with surrounding vehicles and road infrastructure and analyse lane changes, mimicking the behaviour of a swarm of fish. In the future, this could be integrated with standard swarming business models in platoons or more sophisticated business models for logistics, considering delivery speed and accommodating emergency transports, for instance, when production line disruptions pose a risk in the petrochemical or assembly industry. Ideally, autonomous swarm solutions comprise self-driving vehicles capable of operating in all weather conditions, on various types of surfaces, and across the globe.

In the context of urban mobility, Aoyama et al. [6] identify subscription-style, 'X-as-a-service' models (for example, Mobility as a Service (MaaS)), which are implemented in urban areas and entail the provision of services for compensation. Notably, prior to the widespread adoption of autonomous vehicles, cities have already emerged as pivotal to the proliferation of innovative mobility services, such as ridesharing platforms (for instance, Uber and Lyft), which are anticipated to evolve concurrently with autonomous vehicles through new iterations of the mobility-as-a-service business model.

The literature further delineates an array of business models specifically tailored for distinct vehicle manufacturers. Pérez-Moure et al. [16] analyse digital mobility business models relevant to autonomous vehicles from the following manufacturers: Toyota, Stellantis, and Honda.

Toyota has developed digital solutions for autonomous vehicles in collaboration with various third-party partners, thereby establishing a comprehensive ecosystem. Central to Toyota's business model is a digital platform known as 'e-palette', which emphasises the autonomous transportation of both people and goods [16]. Services, including door-to-door transport, are accessible through mobile applications [92]. Toyota's partnership with Pony.ai has resulted in the creation of a virtual assistant for this self-driving vehicle [16]. Pérez-Moure et al. [16] further emphasise additional Toyota business models related to mobility services, with a particular focus on sustainability and connectivity.

Subscribers to Toyota's public charging network can access the Toyota Public Charging Network through service apps. Navigation systems can display a list of the nearest public charging stations and their locations. Alternatively, smartphone apps such as Zap-Map [99], allow users to search for public charging points, plan longer journeys, pay on participating networks, and share updates with other electric vehicle drivers.

Toyota has adopted advanced business models centred on connectivity. The manufacturer provides a comprehensive connectivity solution that incorporates a variety of services, including authentication systems, mapping services, virtual assistants, and automated adjustments of the vehicle's interior settings. Furthermore, encryption via smartphones is available, along with access to emergency services, such as route guidance and driving recommendations. The manufacturer has been developing its digital platform in collaboration with Amazon as a technology partner, utilising diverse devices for data storage, which Toyota processes to offer services directly to customers and to facilitate the extension of third-party offers via the platform. These services include personalised maintenance plans, flexible leasing options, and on-demand insurance, in partnership with the MS&AD Insurance Group's Aioi Nissay Dowa Insurance Company [91]. Additionally, Toyota offers a car-sharing service while retaining ownership of the vehicles in its fleet. Access to these services is available through an app developed in conjunction with Ridecell [93], designed to facilitate on-demand services.

Stellantis has adopted a business model pertaining to autonomous vehicles that closely resembles that of Toyota. The manufacturer has entered into a contractual agreement for cooperation and collaboration with a partner to deploy a fleet of autonomous light commercial vehicles that will provide door-to-door services by the year 2025 [16].

Stellantis employs a business model that is centred on autonomous vehicles (AVs), which markedly differs from models associated with sustainability and connectivity, akin to the strategy embraced by Toyota.

The business model related to sustainability encompasses the enhancement of charging options and the accelerated implementation of smart grid technology [16]. Stellantis manages a network of charging stations designed with exclusive conditions for its clientele, covering over 15,000 locations, which includes 2 million stations located in urban centres. In alignment with its commitment to electric vehicles, Stellantis has, through its collaboration with Kiri, introduced a system that incentivises drivers for demonstrating sustainable driving habits via 'eCoins', which can be utilised in the Kiri marketplace. These eCoins are accumulated through a service tool application, in which driving efficiency is evaluated [16].

Stellantis has adopted business models related to connectivity. The manufacturer is pursuing a software development strategy that leverages its technical skills and ability to recruit talent to enhance expertise, thereby augmenting its existing development resources [16]. In collaboration with Amazon, Stellantis has launched customer-centric connected experiences across the brand's vehicles through applications that offer a comprehensive range of services, including trip planning, voice assistance, payment services, and Amazon's virtual assistant [16]. Drivers can choose from these applications available through connectivity and make payments via a subscription model. Furthermore, Stellantis structures its digital platform business models on Amazon's digital infrastructure.

According to Lampón and Muñoz-Dueñas [16], in the context of services related to sustainable mobility, Honda has developed a portable battery system known as the Mobile Power Pack [16]. A mobile application facilitates the monitoring of charging points, thereby eliminating the necessity to accompany the vehicle.

Honda provides voice assistant services, navigation, and various applications from multiple providers, including Apple and Amazon, accessible through a mobile application. Additionally, users can utilise this application to locate their vehicles, unlock them, or check their statuses. Honda collaborates with HINO to enhance the value of its Mobility as a Service (Maas) business and improve mobility service levels for users [16]. Although numerous definitions exist, 'Maas' can be defined as 'a transport distribution concept' focused on passenger transport, combining a range of transport modes and services to provide a user-oriented transport solution via a single interface.' [100].

With regard to business models related to autonomous vehicles, Honda launched a Cruise AV testing programme in Japan in 2021 [42]. This vehicle began testing in Japan that same year and was primarily focused on improving passenger mobility. At the same time, the company is developing another autonomous vehicle that centres on shared mobility.

Cabanelas et al. [16] present a classification of the new business model typology, founded on Product-Service-System ('PSS') categorisation, alongside the capabilities definition as outlined by SAE [16]. They emphasise the significance of digital platforms and mobile applications in the development of business models centred around ridesharing or vehicle-sharing, which involves monetary compensation and a corresponding reduction in control over the mobility asset. In this scenario, the owner (primary user) bears responsibility for all fixed and sunk costs (e.g., depreciation, repair, maintenance, insurance), whilst expressing a desire to distribute these fixed costs among other travellers by engaging in the sharing economy through platforms or applications. Ultimately, these business models propose new services, including insurance.

Autonomous vehicles represent a significant component in the advancement of digital mobility business models. Firstly, they offer an opportunity to develop business models predicated on the characteristic of flexible utilisation of these vehicles, which can operate independently as well as in groups consisting of both semi-autonomous and fully autonomous vehicles. In this manner, it becomes feasible to customise digital mobility services to meet the diverse needs of various customers, whether individual or collective. Secondly, the implementation of autonomous vehicles is anticipated to improve road safety. The aforementioned advantages of autonomous vehicles are projected to foster increased public confidence in their usage, and, as a result, stimulate heightened interest in digital mobility services provided through autonomous vehicles.

4.3 Definition of Digital Mobility Services and Description of the Types of Such Services

To conduct the analyses presented in this monograph, it is essential to establish a clear definition of the term 'digital mobility service'. The following preliminary observations should be noted in the process of defining this term. Firstly, it is important to recognise that the concept lacks a definition in current legislation. Secondly, the term is found in academic literature in contexts pertaining to the utilisation of both private and public transportation. Thirdly, the scope of this definition will be strictly confined to wheeled transportation, thus excluding rail, maritime, and aerial transport.

Digital mobility services encompass various service categories, including:

(1) access to transportation services provided by third-party entities using their vehicles,
(2) access to vehicles to facilitate journeys with such vehicles,
(3) assistance for drivers in undertaking tasks directly related to journey planning and execution,
(4) access to content (including voice and data) that, while non-essential for journey planning and execution, enhances the overall comfort of the experience,
(5) support for managing both passenger and freight transportation.

The services are characterised by the following attributes:

(1) They are delivered via the internet;
(2) They operate within an ecosystem;
(3) They are typically provided for compensation;
(4) They are intended for an adult audience;
(5) They do not constitute essential equipment for the vehicle;
(6) They are exclusively dedicated to autonomous vehicles, thereby functioning as a mode of transportation.

A comprehensive elucidation of these service categories is provided below. Additionally, the legal qualifications pertinent to these services are delineated in Sect. 4.4.

It is essential to recognise that the distinction outlined in the previous definition of remuneration is substantiated as follows. The criterion 'normally provided for remuneration' aligns with Article 57 of the Treaty on the Functioning of the European Union ('TFEU') [19], which stipulates that services governed by the TFEU are generally provided for remuneration. Considering the analogous nature of these concepts, the case law of the Court of Justice of the European Union ('CJEU') [15] regarding Article 57 TFEU holds substantial significance. Within this body of case law, the CJEU has interpreted the concept of remuneration broadly, encompassing any advantage that serves as consideration for the service rendered. The remuneration criterion indicates that the service must represent an economic activity, as opposed to social services of a non-economic nature typically provided by the State.

Within the definition of Information Society Services ('ISS'), which also includes the phrase 'normally provided for remuneration', the CJEU has noted that the recipient of the service need not necessarily provide remuneration directly but may instead generate income through 'advertisements posted on a website'. This body of case law suggests that the provision of personal data or funding through advertising—common in the context of OTT services—would qualify as a relevant economic consideration. Consequently, the criterion 'normally provided for remuneration' is likely to be satisfied for OTT services in practical applications (the term 'OTT services' was defined in Sect. 4.4).

This definition delineates digital mobility services from a material perspective. It distinguishes five categories of digital mobility services and six application objectives. These objectives facilitate the identification of the intended audience for these services. The recipients of digital mobility services have been classified based on the purpose of these services. The intended utilisation of the services for the relevant groups of recipients allows for the conclusion that the services generate added value for both the recipients and the service providers. This definition delineates three groups of recipients of digital mobility services.

The initial category of digital mobility services concerns the provision of transportation services executed by a third-party employing its vehicle. A notable example of such 'door-to-door' services is represented by Toyota's offerings, which are rendered in collaboration with the partner Pony.ai through a mobile application referred to as 'e-palette'. Within the framework of this service, a user may request medical services via the e-palette application, at which point the autonomous medical vehicle is dispatched to their residence (refer to Sect. 4.2 for further details).

The second category of digital mobility services encompasses digital services that furnish access to a mode of transportation, such as a vehicle. This category comprises car-sharing services. According to Münzel et al. [16], this category of services represents peer-to-peer ('P2P') business models, whereby individuals offer their private vehicles for rental through digital platforms. Bergman et al. [10] contend that the transition from product-based to service-oriented approaches, along with the integration of information technology into mobility, facilitates the emergence of various forms of car-sharing, including car clubs that provide short-term membership-based rentals offered by non-profit organisations or for-profit enterprises. Conversely, Svennevik et al. [16] assert that car-sharing entails the incorporation of novel technologies for accessing vehicles via digital platforms and integrated information technology. Concurrently, numerous material aspects of car-sharing remain analogous to those within the current mobility system, such as vehicles, roads, and parking infrastructure. The transformation pertains more to communication rather than to physical artefacts and infrastructures; the principal modifications lie in internet access, the utilisation of software and hardware devices, including smartphones for maps, communication, keyless technologies, and payment technologies, among others. The significance of digital platforms is further elaborated upon in the discourse regarding the other types of services that constitute digital mobility services.

Digital platforms, also known as platform companies, are entities that create digital communities and marketplaces to facilitate interaction and transactions [16].

Although there is no universally accepted definition of a platform, most descriptions emphasise the concept that platforms utilise digital technology to enable interaction between users and providers [16].

The status symbol of vehicles within society is diminishing. The preferences of emerging generations of consumers are shifting from vehicle ownership to utilising the functionalities offered by such vehicles. Consumers seek to engage with vehicles for individual mobility while avoiding the drawbacks associated with ownership, including depreciation, parking, insurance, inspection, and repair costs [16]. The implementation of autonomous vehicles equipped with level 4 and 5 driving automation will further mitigate the expenses associated with employing human drivers as taxi operators. Genzlinger et al. [16] indicate that mobile services related to car sharing may undermine the foundational business of manufacturers—the sale of vehicles—by reducing the demand for new automobiles. In collaboration with technology partners providing digital platforms for mobility services, the role of vehicle manufacturers may evolve to that of mere providers of transportation means. Moreover, it is essential to consider that vehicle manufacturers are forfeiting direct engagement with consumers to technology partners managing digital platforms.

The third category of digital mobility services pertains to vehicle operators. Digital mobility services aimed at operators are classified based on criteria for selecting activities deemed essential for planning and executing a journey. This category specifically addresses activities that assist the operator in the planning and execution of a journey using a vehicle.

The examination of digital mobility services in relation to autonomous vehicles should clarify the following dependencies. The services associated with the third category of digital mobility services are linked to the vehicle's level of automation. As this level increases, the artificial agent takes on a more significant role in driving tasks. Furthermore, the duration for which these functions are carried out, in relation to the time the vehicle is operational, also extends. In other words, as the degree of automation rises, the period during which a human driver can take on the role of a passenger decreases. Consequently, the third category of services becomes irrelevant for such individuals.

Services related to the third category of digital mobility services also apply to vehicles operating in organised groups, commonly referred to as 'platooning'. The business models outlined in Sect. 4.2 assume that, in these cases, vehicles travel along the roadway in columns, guided by a lead vehicle that is an integral part of such a formation. It is essential to distinguish between two scenarios in which a vehicle regulates a column of vehicles. In the first scenario, a human operator is present within the autonomous control vehicle, requiring that the operator regain control of the vehicle upon the request of the Automated Driving System ('ADS'). The operator remains responsible for intervening when prompted by the ADS. While the operator may safely divert their attention from driving responsibilities, they must be prepared to intervene within a specified timeframe if alerted by the automated system or when a vehicle detaches from the platoon. In the second scenario, no human operator is responsible for controlling the vehicle.

The fourth category of digital mobility services pertains to both the driver and the passengers of the vehicle. These services do not directly relate to the planning and execution of the journey; instead, they function as information services that offer entertainment for both the driver and passengers. Consequently, even when the vehicle operates in autonomous mode, meaning without a human driver, a human passenger continues to be a potential recipient of this type of service.

The fifth category of digital mobility services focuses on achieving benefits related to flexibility and utilisation efficiency. The intended recipients of this type of service are the owners or operators of profit-making vehicle fleets. The individuals inside the vehicle, including both the driver and the passenger, as well as the person requesting the transportation service, do not form the direct recipients of this service.

The provision of the fifth category of digital mobility services exerts a positive influence on the recipients of the other categories of these services. The management of vehicle fleets enhances the interest in the first category of services, as it allows for the assumption that the transportation service sought by this recipient will be delivered with superior quality parameters (for instance, by improving the punctuality of the arrival and departure of transportation means). Furthermore, the provision of a fifth category of digital mobility service will also impact drivers and passengers, who are the recipients of the second and third categories of digital mobility services. Fleet management elevates the quality of vehicle utilisation by drivers.

The provision of a category of digital mobility services has the potential to enhance interest in other types of digital mobility services for several reasons. It serves to augment the interest of the recipient of the initial type of digital mobility service, as it enables an assumption that the transportation service to which this recipient seeks access will be delivered with superior quality (e.g., regarding the punctuality of the arrival and departure of the means of transport).

Digital mobility services should pertain to motor vehicles, including autonomous vehicles. The motor vehicle signifies the initial step in the transition from a product-centric to a service-centric paradigm. A motor vehicle must be comprehended from the perspective of its transportation functions, which the digital mobility services reference. Consequently, the vehicle is regarded as a means of mobility rather than as a collector's item. The utility value of the vehicle necessitates perceiving it in terms of motion.

The movement of individuals and goods is integral to achieving various societal objectives. Given the distinct recipients of digital mobility services, it is essential to view society from the perspective of its individual constituents and the goals they aspire to reach, while also recognising the necessity of utilising transport methods. The relationship between digital mobility services and autonomous vehicles can be understood in three dimensions: the request for transport by the vehicle, the necessity for transport by the vehicle, and the provision of mobility by the vehicle.

In this third instance, digital support services can effectively address challenges such as modifying road infrastructure to enable the cessation of vehicle movement after their journeys. This includes services related to the management of urban parking facilities, along with the parking provisions designated for these vehicles.

Digital mobility services are developed and implemented within a framework that integrates the use of digital tools alongside the various stakeholders involved in creating or co-creating the components of this service. The interactions among the stakeholders engaged in the development of a digital mobility service are analysed in greater detail in Sect. 4.1.

The provision of mobility services requires resources from both the customer seeking digital mobility services and the recipient of those services. The Internet acts as the transmission medium for these services, thus necessitating that both the requester and the provider have independent Internet access. The requester, in particular, should have mobile Internet access to effectively utilise the services in a range of circumstances beyond fixed locations.

With respect to the utilisation of digital platform services, Steinberg [16] presents the following scenario: 'Imagine a familiar scene: a ride-seeker takes out their smartphone, opens the Uber app, and orders a ride. A few minutes later, an Uber driver picks up a passenger in a Toyota Prius, driving them to their destination.' This process is governed by a data-driven application equipped with algorithms that optimise routes, manage driver performance, and collect data regarding the destination.

Nevertheless, the resources of the provider for digital mobility services require further enhancement. A critical resource is the accessibility of digital platforms. Hodson et al. [16] observe that while ride-hailing platforms such as Uber are the most recognised, the spectrum of digital mobility platforms is becoming increasingly diverse. Micro-mobility platforms for shared bicycles (e.g., Beryl; Nextbike) and e-scooters (e.g., Voi; Neuron) operate in numerous cities worldwide, alongside a variety of ticketing systems (e.g., Masabi; Vix Technology) and journey-planning platforms (e.g., Citymapper; Moovit). Moreover, 'mobility-as-a-service' platforms integrate journey planning, payment, and ticketing across different modes of transport (e.g., Trafi; Maas Global). Caprotti et al. [16] indicate that platforms function as 'digitally-enabled assemblages' that facilitate and deliver services (e.g., Uber for ride-hailing, Deliveroo for meal delivery), allowing for effortless initiation and scaling in various towns and cities. Furthermore, Schüßler et al. [16] assert that there is a collective understanding that digital platforms signify an advancement in the dismantling of organisational boundaries instigated by the Internet and communication technologies.

The use of digital platforms requires a thorough understanding of the context in which they operate, including urban infrastructure. Hodson et al. [16] have noted that bike-sharing platforms significantly depend on existing road and cycling facilities, which include designated docking stations along main roads and at public transport hubs. Moreover, the development of front-end applications, back-end systems, and the bicycles themselves is crucial. These platforms not only utilise existing urban mobility resources but also have the potential to improve current services (for instance, by placing bike docking stations at train stations to facilitate 'last mile' connectivity) or to compete with current options (such as certain ride-hailing services).

The integration of contemporary systems with these platforms engenders complex interactions in which they 'fuse' or 'hybridise' in innovative manners, influenced by diverse social interests. This evolution poses challenges to the established institutional frameworks of existing systems [16].

In evaluating the significance of the digital platform as a resource for providers of digital mobility services, it is essential to clarify the meaning of this term. According to Cusumano [16], the term 'platform' gained widespread use in management research to denote a foundation of components around which an organisation develops a related yet distinct set of products (or services). Negoro et al. [16] characterise this model as a core technology shared by numerous digital products, including display technology, automotive chassis, printed circuits in the equipment of autonomous vehicles, and so forth. This definition is consistent with a parallel one posited by Gawer et al. [16], wherein the platform is described as 'a core technological building block upon which organisations build complementary technologies, products or services'. Bratton [12] refers to the stack as a 'modular and interdependent vertical order' and 'a kind of platform that also happens to be structured through vertically interoperable layers'. He identifies the stack as one of the earliest industrial models of the platform. As noted by Cusumano [16], one of the fundamental referents of the stack model of the platform is the automobile chassis, commonly referred to as the platform. The implementation of a base platform for various car models represents a practice that intertwines base-level standardisation with maximal second-level variation [16].

The second variant of the platform concept is the intermediary model. Unlike the layered model, this configuration represents a horizontal structure of the platform, functioning as a mediatory device that facilitates third-party transactions. A credit card exemplifies one of its forms [16], while Airbnb and Uber serve as extensions of this model, acting as intermediaries between room hosts and renters, as well as between drivers and riders, thereby creating multisided markets [16]. Most economic literature pertaining to platforms emphasises their role in connecting two or more market 'sides' to enable transactions—whether through a credit card or the Uber application. The second usage aligns with what [16] designates as an 'industry platform', which relies on complements and network effects to bestow value upon the service—illustrated by a computer possessing relatively minimal value without the applications that operate on it and the users who depend on such applications. Platforms function as intermediaries that facilitate multisided markets: 'Essentially, [platforms] are a newly predominant type of business model premised upon bringing different groups together' [16]. Proposed as an alternative to the competing terms 'gig economy' or 'fourth industrial revolution', platform capitalism serves as a descriptor for 'the effects of digital technology on capitalism' [16].

The provision of digital mobility services necessitates processing substantial volumes of data, rendering these services reliant not only on connectivity but also on customer-identifying information. From the perspective of the digital mobility service provider, managing significant data volumes requires either in-house resources for data storage or third-party cloud services. However, this data generates additional value, offering considerable insights to the service provider and its collaborators (including vehicle manufacturers) regarding customer behaviour and preferences. This understanding, among other factors, facilitates the delivery of customised (personalised) advertising. Sadowski [16] employs the term 'Commercialisation of Driving Data', positing that 'data is capital'. In alignment with this perspective, it is essential to acknowledge that the provision of digital mobility services yields dual value—benefiting not only the service providers but also the customers (as indicated previously) and third parties (e.g., vehicle manufacturers). In this context, Krafft et al. [16] assert that, in accordance with shifts in the types and forms of data that manufacturers collect, the purposes for utilising such data have similarly evolved. Specifically, the traditional focus of manufacturers on interactions with retailers and channel intermediaries has transformed to include end consumers. Recently, while automobile manufacturers have utilised their network of licensed dealers to sell vehicles, they have also begun to engage in direct-to-consumer communication (beyond dealer interactions) through contemporary channels and media. Additionally, manufacturers such as Porsche, BMW, and Audi are capitalising on direct-to-consumer communications by developing innovative business models such as the monthly car subscription service.

The processing of extensive volumes of data concerning the provision of digital mobility services is necessitated by the intricate nature of autonomous vehicles. The literature underscores that, with the advent of Connected and Autonomous Vehicles ('CAV's), advancements in automobile automation (e.g., [11, 13]), machine learning (e.g., [16, 16]), autonomous vehicle testing (e.g., [101]), and the policy ramifications associated with self-driving cars (e.g., [16]) further broaden the concept of the vehicle as a medium of social interaction or as an inherently "smart" object. Additional considerations have been raised regarding the necessity for CAVS to implement default data collection and processing protocols, often at the expense of individual privacy [16], and the ramifications for 'future users' of CAVs, who may find themselves either empowered or marginalised by these technological advancements [102]. According to [16], the vehicle merely transforms into another apparatus for the acquisition of personal data. Moreover, as long as driving data remains pivotal for the training of machine learning models, such data will continue to be commercialised and segregated. A notable instance is observed within the insurance industry, in which various devices are employed to curate personalised products [16]. Additional instances pertain to the privacy concerns associated with vehicles equipped with multiple external cameras [16], presenting significant challenges to data protection regulations, particularly in densely populated locales (i.e., urban platforms), as they capture footage of other road users without obtaining their consent (i.e., mobile sensing platforms). Consequently, the processing of substantial volumes of data related to digital mobility services is substantiated by the complexities inherent to autonomous vehicles and the corresponding spectrum of services, aligned with the categories of services delineated in the definition of 'digital mobility services'. In this regard, digital platforms serve as effective instruments, taking into account the intermediary model previously referenced. Within this framework, Hind et al. [16] contend that platformisation exemplifies a significantly broader and contemporary trend within the automotive sector, with automation and datafication being fundamentally interdependent.

4.4 Legal Nature of Digital Mobility Services

The legal nature of digital mobility services, as delineated in Sect. 4.3, will be expounded upon in the subsequent discussion. Digital mobility services are defined as those facilitating transportation objectives at various stages of engagement, specifically aimed at individuals who are planning a journey, operating a vehicle, or managing a fleet. These services are scrutinised within the context of the deployment of autonomous vehicles.

The legal definition of 'digital mobility services' should be established in accordance with the specific categories of this service, as delineated in Sect. 4.3.

The initial category of digital mobility services involves the coordination of transportation for individuals or goods by a third party via a website or mobile application. A ride seeker expresses their interest in transporting either persons or goods over a specific distance and timeframe. Selecting a transport offer available online leads to the establishment of a contract. This contract may be formed between distinct parties, specifically the driver and the customer (ride seeker). In cases where the transportation of goods or individuals is facilitated by an autonomous vehicle, adhering to levels 4 and 5 of SAE [16] automation (i.e., without human intervention), it is not feasible to conclude a contract of carriage with an artificial agent, as it lacks a separate legal identity. Consequently, the customer would enter into a contract of carriage with the carrier operating autonomous vehicles. Should the transportation service occur using buses or coaches, Regulation (EU) No 181/2011 [16] is applicable to safeguard passenger rights.

For instance, under Polish law, the contract of carriage is governed by the provisions outlined in Articles 774–793 of the Polish Civil Code (1964) [5] and the Transport Law (1984) [2]. This contract typically constitutes an adhesion contract, thus prohibiting the possibility of individually negotiating its terms. The contract encompasses the following objectively essential elements: the route of carriage, remuneration for the carrier (noting that free transport is permissible), and the subject matter of the transport. The subject of the contract of carriage may pertain to either individuals or goods. Consequently, the conclusion of the contract is marked by the submission and acceptance of the offer. The acceptance process may occur in multiple stages. FLIX, in point 3.3.3 of the General Terms and Conditions of Carriage, states: 'It should be noted that the statutory provisions regarding cancellation rights in the case of distance contracts shall not apply as far as contracts of carriage concluded online by way of distance selling are concerned, and in the case of which the operator undertakes, upon the contract's conclusion, to perform the service at a specific time or within a stipulated period. This contrasts with the online mail-order business. Our conditions concerning cancellation remain, however, unaffected by the above.' [16].

The carrier's obligations comprise ensuring suitable conditions for passengers, specifically concerning safety, hygiene, and comfort; accountability for the luggage transported by the passenger if damage arises from the carrier's wilful misconduct or gross negligence; and responsibility for luggage entrusted directly to them, in accordance with the regulations governing the transport of goods. The contract of carriage for goods entails an obligation to transport such goods for remuneration. The goods to be transported must be detailed in an annex. The contract is concluded upon the acceptance of the shipment for transport and the transfer of the waybill, which provides comprehensive details regarding the sender's and recipient's addresses, destination, shipment marking, including type, quantity, packaging method, and the value of any valuables.

The carrier is authorised to provide a transport service in exchange for remuneration payable in advance via online platforms. Furthermore, it is feasible for a third party to act as an intermediary through a dedicated ride-hailing platform when executing a transport agreement and collecting remuneration in advance for such transport. In this instance, the ride-hailing platform derives revenue from two distinct sources. The first scenario pertains to the coordination of privacy-sensitive customers with available drivers who offer flexible ride-hailing services, while the second scenario relates to targeted advertising based on the information gathered from customers [16]. For instance, UBER delineates its role in this manner, as articulated in point 2.2.2 of the General Terms of Use, stating that it enables 'to enter into direct contracts for transportation with Third Party Providers and/or to settle fares for transportation (together the 'Intermediation Services').' [16]. Concurrently, in point 3.1, first paragraph, of these General Terms, UBER specifies that 'Uber does not provide the Third Party Services. All Third Party Providers are independent third-party contractors, registered businesses, or other legal entities who are not employed by Uber.' Moreover, point 3.1, second paragraph, of the General Terms and Conditions defines UBER as 'disclosed agent between you and the Third Party Providers to enable them to provide the Third Party Services to you.' The functions performed by UBER are carried out through substantial activities, wherein it facilitates the conditions necessary for the conclusion of an agreement by the parties involved.

In Polish law, brokerage has been directly regulated concerning real estate trading since 1 September 2017, as stipulated in Articles 179a and 179b of the Polish Real Estate Management Act (1997) [4]. Real estate brokerage entails the execution of compensated activities aimed at facilitating the conclusion of contracts by other parties: (1) the acquisition or disposal of rights to real estate; (2) the acquisition or disposal of cooperative ownership rights to premises; (3) the lease or tenancy of real estate or parts thereof; (4) any other activities not specified in points 1–3 that pertain to rights to real estate or parts thereof. A brokerage contract cannot be applied in the context of concluding contracts of carriage. Furthermore, the Polish Civil Code regulates the agency contract in Title XXIII (Articles 758–764^9 of the Civil Code). According to the definition of an 'agency contract' outlined in Article 758 of the Civil Code § 1, this is an agreement in which the individual accepting the order (agent) commits, within the framework of their enterprise activities, to consistently mediate, for remuneration, in the execution of agreements with clients on behalf of the entrepreneur issuing the order, or to conclude such agreements on their behalf. It is understood that the materially relevant provisions (essentialia negotii) of the agency contract encompass:

(1) The obligation of the agent within the scope of the activities of his enterprise to consistently mediate in the conclusion of contracts with clients on behalf of the principal, or to consistently conclude contracts with clients while acting on behalf of the principal;
(2) Indication of the subject of the agent's services, specifically the agreements in which he is to mediate or which he is to conclude on behalf of the principal;
(3) The mandator's obligation to compensate the agent's commission.

Regardless of the form of the agency agreement—whether it relates to intermediation or representation—the actions of the agent are grounded in a continuous legal relationship. The agent may consistently mediate in the formation of agreements for clients or may permanently finalise such agreements. The consistency of the agent's actions is reflected in the repeatability and, at times, the continuity of behaviour, which constitutes the fulfilment of the contract. An agreement in which a party commits to mediate or finalise a contract but does not do so on a permanent basis does not qualify as an agency agreement. Instead, it is classified as an agreement on ordinary intermediation, which is currently unregulated, or as a contract of mandate, as indicated in the Judgement of the Court of Appeal in Gdańsk, II AKa 387/15 [16].

In the judgement C-3/04 (2006) [16], the European Court of Justice stated that 'The number of transactions concluded by the intermediary in the name and on behalf of the principal is usually an indicator of the permanent nature of the intermediation. [...] where the intermediary is entrusted with the conclusion of a single contract, which is then extended for several years, the condition of the permanent nature of the intermediation specified in that provision requires the principal to entrust the intermediary with the intermediation of subsequent extensions of that contract.' Consequently, it is challenging to identify an agency agreement between an individual interested in utilising the transport service and the entity managing the digital platform. Such a legal relationship may exist between the driver and the entity operating the platform, provided that the permanence of that legal relationship can be established. Regarding the individual interested in using the transport services and the entity operating the digital platform, there exists a legal relationship wherein that individual requests, through electronic means of communication, the identification of available drivers in the vicinity willing to perform the transport service.

The digital platform enables payment transfer to the driver in cases of advance payment, akin to the model used by UBER. Consequently, there is a specific set of obligations for the entity managing the digital platform towards the individual seeking transportation services. These obligations include connecting the ride seeker with a driver willing to undertake the journey and ensuring the transfer of payment for this ride to the driver. Furthermore, UBER specifies in section 3.1 of the General Terms and Conditions that it acts as a disclosed agent between the ride seeker and the third-party provider of transportation services. In accordance with the Polish Civil Code, the services provided by the entity operating the digital platform may be defined as services not governed by other provisions of the Polish Civil Code (Article 750 of the Polish Civil Code). The regulations concerning orders apply correspondingly to these provisions. Such an agreement is established through facta concludentia.

A distinct legal relationship formed between the individual seeking transportation services and another party interested in executing this transport service constitutes the contract of carriage.

The activities undertaken by the digital platform operator can be characterised as 'provision of services by electronic means' as delineated in Article 2(4) of the Polish Act on the Provision of Services by Electronic Means (2002) [3]. This classification pertains to the execution of a service rendered without the simultaneous presence of the parties (remotely), through the transmission of data at the individual request of the service recipient, which is conveyed and received via electronic processing devices, including digital compression and data storage. All of this is transmitted, received, or conveyed entirely via a telecommunications network as defined by the Polish Electronic Communications Law (2024) [1].

The second category of digital mobility services, outlined in Sect. 4.2, involves the requisition of a motor vehicle for the purpose of undertaking a journey. Within the scope of this service, it is presumed that the vehicle will operate in autonomous driving modes, allowing for the possibility of human operation; this excludes driving automation modes 4 and 5. Conversely, if the vehicle were to revert to solely human operation, the individual would simply take on the role of a passenger, thereby categorising the service as a type of the first category of digital mobility services.

In Sect. 5.1, peer-to-peer (P2P) business models are examined, specifically focusing on individuals offering their personal vehicles for rental through online platforms. Within one of these business models, the digital platform operator facilitates the connection between individuals seeking to rent a vehicle and those offering such vehicles for rent. The classification of the legal relationship between the prospective lessee and the lessor is analogous to that which exists when the owner of a fleet of vehicles directly offers them for rent via a website. The Polish Civil Code governs leasing under Title XVII, Section I, Chapter I (Articles 659–579). Pursuant to Article 659 § 1 of the Polish Civil Code, the lease contract pertains to a movable item provided to the lessee for use in exchange for remuneration for a specified or unspecified duration.

The third category of digital mobility services pertains to assisting drivers in activities directly associated with the planning and execution of a trip. This service is specifically designed for individuals who are responsible for planning or executing a journey. Assistance is rendered through mobile applications, the quantity of which is contingent upon the demand for services that facilitate the planning and implementation of travel. Illustrative examples of applications pertaining to trip planning include those that determine optimal routes, estimate travel durations, or outline fees applicable to toll motorways, tunnels, or ferry crossings. Additionally, such applications may indicate the locations of charging stations for electric vehicles along the designated route, particularly when the autonomous vehicle in question is also an electric vehicle.

This support is facilitated through mobile applications, the number of which depends on the demand for services that assist in travel planning and execution. Examples include applications that provide travel time estimates for routes, as well as fees associated with the use of toll motorway sections, tunnels, or ferry crossings. Furthermore, these applications may also indicate the locations of electric vehicle charging stations along the route, particularly when the autonomous vehicle in question is electric.

Applications may also relate to the monitoring of parameters associated with the sensitive oversight of critical vehicle components that ensure safe driving. According to Sumail et al. [16], maintenance applications are installed on mobile devices, allowing users to retrieve data from the vehicle using the onboard diagnostics (OBD) protocol. The primary objective is to collect and disseminate internal vehicle diagnostics. Contemporary vehicles are equipped with an electronic control unit (ECU) responsible for sensing environmental conditions and regulating vehicle functionalities [16]. As noted by Sumaila et al. [16], car maintenance kits typically include a mobile application—either from the vehicle manufacturer or a compatible application available on Google Play or the App Store. This application facilitates the connection between the mobile device and the OBD-II dongle via Bluetooth or Wi-Fi connections (i.e., all applications included in this study use Bluetooth connections). Once the dongle and mobile device have successfully connected, existing users must log into the application; otherwise, new users are required to create an account. The user configures the application by providing essential information regarding the vehicle, such as make, model, year, current odometer reading, and engine size. Subsequently, the application begins monitoring and transmitting vehicle information, including acceleration, speed, GPS data, and any fault codes that may indicate a malfunctioning vehicle condition. This information is displayed to the driver through the application's dashboard or the mobile device.

The fourth category of digital mobility services pertains to access to content that is not directly associated with the planning and execution of a journey; rather, it is designed to enhance travel comfort. This encompasses content that provides entertainment, which, while not instrumental to journey planning or vehicle operation, significantly contributes to the overall quality of the travel experience. Services within this fourth category are also accessible via mobile applications. In contrast to the services classified under the third category, those of the fourth category are additionally available for passengers. Consequently, it is possible to utilise the fourth type of digital mobility services, relating to the fourth and fifth levels of automation, wherein the artificial agent undertakes the entirety of the driving task.

The applications related to the fourth and fifth categories of digital mobility services may be available as mobile applications or as software integrated within the vehicle. Applications that are accessible within the vehicle make up its infotainment system. Mandal et al. [16] assert that: 'with the increasing demand for more connected vehicles, and widespread use of smartphones, in-car entertainment is getting more and more sophisticated'. Software manufacturers, namely Google and Apple, have introduced infotainment systems (Apple CarPlay and Android Auto). Due to their widespread popularity, vehicle manufacturers have begun the practice of adopting these systems, thereby phasing out their proprietary infotainment solutions. An example of such proprietary systems is Ford's infotainment system, SYNC.

The fifth category of digital mobility services is also accessible through mobile applications. These services are designed to assist fleet operators in guaranteeing flexibility and efficiency within transportation. Consequently, the observations provided below regarding the legal nature of services will apply collectively to the third, fourth, and fifth categories of digital mobility services.

Mobile applications are intangible goods primarily acquired from online repositories. They are categorised under the software domain, alongside operating systems and middleware [16]. Specifically, they consist of computer programmes designed with functionality tailored to execute particular tasks for end-users, which can be made available either locally on a device or remotely [16]. Applications may be delivered as digital content, which can be integrated into products, downloaded (for instance, an application), or streamed online (for example, music) [16]. Article 2(12) of Directive 2011/83/EU [16] delineates the term 'digital content'. However, the legislator mandates that the interpretation of this term be established through reference to Article 2(1) of Directive (EU) 2019/770 [16]. According to this definition, digital content is considered data that is produced and supplied in digital form. This definition encompasses applications, with the exceptions specified in Article 3(4), 5(g)-(h). Furthermore, Article 2(16) of Directive 2011/83/EU [16], which pertains to the understanding of the concept of 'digital service', references the definition provided in Article 2(2) of Directive (EU) 2019/770 [16]. Per this definition, 'digital service' signifies: (a) a service that enables the consumer to create, process, store, or access data in digital form, or (b) a service that facilitates the sharing or interaction with data in digital form, which has been uploaded or created by the consumer or other users of that service. As articulated in recital (19), Directive (EU) 2019/770 [16] encompasses computer programmes, applications, video files, audio files, music files, digital games, e-books, or other e-publications, as well as digital services that support the creation, processing, access, or storage of data in digital form. This encapsulates software as a service, including video and audio sharing, file hosting, word processing, or gaming services provided within cloud computing environments and social media.

According to Recital 21 of Directive (EU) 2019/770 [16], digital content or a digital service that is incorporated into or interconnected with goods in such a manner should fall within the purview of Directive (EU) 2019/771 [16] if it is provided alongside the goods under a sales contract pertaining to those goods. The provision of the incorporated or interconnected digital content or service constitutes an integral part of the sales contract with the seller. In Article 2(5)(b) of Directive (EU) 2019/771 [16], the term 'goods' encompasses any tangible movable items that incorporate or are interconnected with digital content or a digital service to the extent that the absence of such digital content or digital service would inhibit the goods from performing their intended functions (referred to as 'goods with digital elements').

Lippoldt et al. [16] defined 'embedded software' as software that is permanently integrated into a specific hardware unit. This software serves as an essential component of the product within which it is permanently situated. Unlike other software types, embedded software is marketed as an integral part of the hardware product on which it resides. In contemporary economies, embedded software is increasingly prevalent due to its application in a myriad of electronic products and systems, notably including vehicles. Automotive manufacturers extensively utilise embedded software applications. The market for embedded software is experiencing rapid growth, with a significant portion occurring beyond the conventional software production sector. Vehicles equipped with pre-installed embedded applications facilitate the execution of original tasks. These vehicles, including autonomous vehicles, are classified as connected objects. Such an intelligent, connected object is fundamentally based on the Internet of Things ('IoT'). Consequently, an 'autonomous vehicle' is characterised as a 'connected product' under Article 2(5) of Regulation (EU) 2023/2854 ('Data Act') [16], thus falling under the purview of this Regulation.

The classification of a contract for digital services as either a sales contract or a service contract requires a precise definition regarding the type of digital content, categorising it as 'goods' or 'services'. The literature also introduces the notion of a 'sui generis' contract. In the case of a sales contract, the seller transfers ownership of the goods to their contractual partner, the buyer or purchaser. When an application is permanently downloaded onto a mobile device for a fee, a sale of that specific copy of the application has indeed occurred. Regarding computer software downloaded online, the Court of Justice of the European Union (CJEU) articulated that 'the downloading of a copy of a computer program from the website of the right holder to the client's server and the conclusion of a user license agreement for that copy form an indivisible whole which, as a whole, must be classified as a sale' (Judgement, C-128/2011) [43].

Applications are subject to intellectual property rights and copyright protections. They transcend mere software, representing a confluence of software and artistic works. While software forms a fundamental component of an application, it also includes the interface and audiovisual elements.

Access to these services can be achieved through a website or an application. Utilising such services may require the installation of an application on the user's device. An application is defined as software for which the user must acquire a licence. Typically, this licence is granted under a personal, worldwide, non-exclusive, non-transferable, revocable, and non-sublicensable agreement (or sublicense) and pertains to the installation and/or use of mobile applications and/or websites on the user's device, solely for personal use and for the individual's access to and use of information provided through mobile applications and/or websites. For instance, UBER employs this type of licensing [16].

Digital mobility services generally do not qualify as electronic communications services (ECS). Likewise, digital platform services that facilitate their provision do not qualify as ECS. However, various services associated with digital mobility services may also be delivered through digital platforms, which would indeed qualify as ECS. This conclusion arises from the following considerations. The European Electronic Communications Code (EECC) [24] delineates in Article 2(4) an 'electronic communications service' as services typically offered for remuneration through electronic communications networks (ECNs), which includes the following categories of services: (1) internet access service (IAS); (2) interpersonal communications services (ICS); and (3) services that chiefly consist of the transmission of signals, such as transmission services utilised for machine-to-machine communications and broadcasting.

Interpersonal Communications Services (ICS) are defined under Article 2 of the European Electronic Communications Code (EECC) as a service 'normally provided for remuneration that enables direct interpersonal and interactive exchange of information via electronic communications networks between a finite number of persons, whereby the persons initiating or participating in the communication determine its recipient(s) and does not include services which enable interpersonal and interactive communication merely as a minor ancillary feature that is intrinsically linked to another service.' ICSs encompass two types of services: number-based ICS (NB-ICS) and number-independent ICS (NI-ICS). The distinction between NB-ICSs and NI-ICSS lies in the former's connection with publicly assigned numbering resources, whereas the latter does not utilise such resources. The concept of 'minor ancillary feature' includes, for instance, a communication channel within online games, determined by the features of the communication capabilities of the service. ICSs may be accessed on a standalone basis; however, in certain Member States, they are also commonly marketed in conjunction with other services.

Regarding the various types of digital mobility services, specifically those outlined in points (1) and (2), the function of the internet portal is to facilitate access to services between the customer and the digital mobility services provider. The legal relationship encompasses three categories of entities: internet service providers, which deliver the physical connection between platform operators and end users. The internet service provider ('ISP') supplies a connection that traverses the ECS markets, with ISPs serving as intermediaries. ISPs can be further categorised into three types: (1) Internet Access Providers ('IAP's), which act as ISPs for end users within the 'retail internet access markets'; (2) Connectivity Providers ('CPS'), which provide services to Connectivity Application Providers ('CAP's) within the 'Internet connectivity markets'; and (3) ISPs that engage with one another within the 'wholesale interconnection markets'. Consequently, the internet platform operator, the customer, and the digital mobility services provider all require access to the internet network. Given that the internet access services constitute an ECS, the customer is entitled to the rights specified in the Electronic Communications Code ('EECC'). In its report dated 10 June 2021 [8], the Body of European Regulators for Electronic Communications ('BEREC') categorised services such as social networks, online intermediation (including payment and identification), and advertising services as non ECS services.

Earlier, BEREC (2016) [9] defined the services in question as Over-The-Top ('OTT') services, characterised as 'content, a service or an application that is provided to the end user over the public Internet'. By including in the definition that the provision may encompass either content, a service, or an application, it implies that any offering delivered over the public Internet qualifies as an OTT service. BEREC further presents the following taxonomy:

(1) OTT-0: an Over-the-Top (OTT) service that qualifies as an Electronic Communications Service (ECS);

(2) OTT-1: an OTT service that does not qualify as an ECS but potentially competes with an ECS;
(3) OTT-2: other OTT services. This taxonomy is depicted in the figure below.
(4) OTT-2 services are defined as those services provided over the Internet that do not compete with ECSs.

The OTT-2 services represent the remaining category, comprising OTT services that are neither classified as ECS nor possess the potential to compete with ECSs. It is pertinent to note that the qualification of an OTT service as ECS and its competitive potential with ECSs are both critical components shaping the taxonomy; however, the determination of whether an OTT service qualifies as ECS is independent of its potential to compete with ECSs. OTT-2 services are delineated as those provided via the Internet that do not engage in competition with ECSs. The breadth of OTT-2 related issues currently under examination is extensive, encompassing social platforms, e-commerce, and search engines.

National regulatory authorities engage in classifying these services on a case-by-case basis. In this context, the authorities evaluate the distinction between ancillary and principal services, as well as the limitations or absence of limitations regarding the number of users it addresses, to differentiate an ICS from a social network. Furthermore, the authorities consistently analyse new services and their appropriate classifications within this ever-evolving sector. Additionally, services may develop over time, leading to potential changes in their classifications.

Numerous regulations, including Regulation (EU) 2016/679 enacted by the European Union (EU), currently exist and impose stringent mandates regarding the use of personal data. To date, four legislative measures have been established to set the benchmark for Europe: Regulation (EU) 2023/2854 ('Data Act'), Regulation (EU) 2022/86 (Data Governance Act), Regulation (EU) 2022/2065 ('DSA') [16], and Regulation (EU) 2022/1925 ('DMA') [16], all of which are anticipated to become effective from September 2025 onward [18].

Nevertheless, when considering the range of digital mobility services related to 3–5 types of these offerings, it cannot be overlooked that these services may concurrently function as interpersonal communication services within the framework of the EECC. This scenario arises when they facilitate an interactive exchange of information (e.g., messaging services or group chats). According to recital 17 of the EECC, 'Interactive communication' denotes that the service permits the recipient of the information to respond.

The introduction of digital mobility services necessitates careful consideration of the legal framework directly associated with autonomous vehicles, including Regulation (EU) 2024/1689 (AI Act, 2024) [16]. The AI Act came into effect on 1 August 2024 and will be implemented in phases over three years. As digital mobility services are examined within the context of autonomous vehicle utilisation, it becomes imperative to refer to the existing regulatory framework governing autonomous vehicles.

Under the current legal framework, autonomous vehicles (AVs) operating within the European Union are governed by a series of regulations. Notably, Regulation

(EU) 2018/858 and Regulation (EU) 2019/2144 specifically relate to automotive standards, covering vehicle approval and market surveillance. Collectively, these regulations establish the type approval framework and are further complemented by international standards, such as those issued by the United Nations Economic Commission for Europe (UNECE), which are incorporated into EU law.

Within the type approval regulations, vehicles must undergo a type approval process to ensure compliance before their introduction to the market. Additional regulations, including Regulation (EU) 2023/988 [16], address broader product safety concerns. Collectively, these regulations guarantee that vehicles conform to EU-wide safety, environmental, and technical standards before entering the market. Although they address traditional safety concerns, which include elements relating to the use of AI, they do not specifically tackle AI-related risks.

The implementation of artificial intelligence (AI) across various industries is governed by legal regulations that, although not specifically designed for autonomous vehicles (AVs), remain applicable. For instance, when AI processes personal data obtained from connected vehicles, Regulation (EU) 2016/679 [16] is relevant, alongside copyright laws that oversee the use of copyrighted content by AI, even in the absence of explicit references to AI. In summary, numerous pre-existing legal frameworks already address the deployment of AI in automotive products and services. Furthermore, the AI Act functions as a specific regulation pertaining to AI technologies. As articulated in Article 3(1) of the AI Act, the definition of 'AI System' is broad and encompasses autonomous vehicles.

In both the European Union and globally, the governance of artificial intelligence (AI) is advancing towards a regulatory framework centred on risk management [16]. This transition is also evident in the AI Act, which delineates four distinct levels of risk. Furthermore, the AI Act advocates for a proactive assessment of the risks that an AI system might present, thereby allowing for an evaluation of these risks prior to the actual deployment of the AI system.

The AI Act stipulates that systems classified as presenting an unacceptable risk will face a prohibition on their use. Such circumstances arise in cases outlined in Article 5 of the AI Act. This prohibition covers the placing of such systems on the market, their introduction into service, or their utilisation. It applies to AI systems characterised by various purposes, including, but not limited to, the use of subliminal techniques that operate beyond an individual's conscious awareness or the employment of intentionally manipulative or deceptive methods. Additionally, it addresses the exploitation of any vulnerabilities associated with natural persons or specific groups, particularly concerning factors such as age, disability, or particular social or economic circumstances. This includes the evaluation or classification of individuals or groups over a designated time frame based on their social behaviours, as well as conducting risk assessments aimed at evaluating or predicting the likelihood of an individual engaging in criminal activity, among other concerns.

Secondly, the AI Act outlines provisions for low-risk AI systems, typically including chatbots and similar applications. It is crucial for users to be explicitly informed that they are interacting with an AI system, and in due course, codes of conduct should be established to guide both developers and users regarding trustworthy AI systems (section 95(1) of the AI Act). AI systems classified as posing minimal risk to individuals or society are expected to follow general guidelines and best practices, without being subjected to additional legal requirements.

Third, the AI Act anticipates high-risk scenarios. Numerous systems used in automated and autonomous vehicles fall into the high-risk category due to their potential implications for public safety and fundamental rights. Consequently, manufacturers of autonomous vehicles and suppliers of in-vehicle software must prepare to meet stringent obligations to ensure compliance with transparency, traceability, data management, security, and liability requirements. Chapter III of the AI Act addresses high-risk AI systems that may have severe adverse consequences in the event of a failure. Such systems are required to meet rigorous criteria regarding data security, transparency, human oversight, and resilience, and they must undergo a mandatory compliance assessment prior to market entry, with ongoing monitoring throughout their life cycle.

Manufacturers of autonomous vehicles and providers of vehicle software should classify artificial intelligence (AI) systems based on their associated risks to ensure compliance with the AI Act. It is vital to assess whether the AI systems used in autonomous vehicles, including but not limited to perception, decision-making, or control systems, could be deemed high-risk under the AI Act. AI is increasingly becoming a vital component of products and services within the automotive industry. In the context of autonomous and driver assistance systems, AI enhances vehicle intelligence, safety, and efficiency by analysing data derived from various driving scenarios to facilitate informed driving decisions, adapt to changing road conditions, and evaluate driver fitness and alertness. Furthermore, AI plays a crucial role in supporting non-safety functions such as media, access, and payment services within vehicles.

The role of AI in regulating vehicle operations has the potential to significantly impact individual rights and freedoms. Failures in the system could lead to accidents, property damage, or personal injury. Upon initial assessment, most AI-based autonomous and assistive systems are likely to be classified as high-risk AI, due to the critical importance of decision-making that occurs within the vehicle (i.e., "on the road"). In contrast, alternative AI applications, such as those related to entertainment or convenience features, are more likely to be classified as low-risk AI.

To comply with industry regulations and standardised EU safety protocols, the AI Act categorises certain AI systems as high-risk AI, depending on their regulation by specific harmonisation legislation (Article 2(2)), which includes frameworks such as type approval and Regulation (EU) 2023/988 [16]. These frameworks require prior conformity assessments and explicit approval. In cases where this applies, the AI Act shall not be enforced directly.

This methodology guarantees that AI applications specific to the automotive sector will predominantly be governed by relevant sectoral legislation, while the high-risk AI provisions within the AI Act will serve as supplementary, albeit overarching, standards. Additionally, the AI Act obliges the EU legislator to amend EU harmonisation law to align it with the high-risk AI standards present within the type approval and GPSR framework (Article 80). Such amendments will incorporate high-risk AI requirements specific to the automotive sector, which may include regulatory sandboxes designed to foster innovation.

Non-regulated AI systems, which are not subject to harmonised EU legislation, must continue to be evaluated in accordance with the stipulations of the AI Act and undergo standard risk assessments. The details of the revised AI-related type approval framework and Regulation (EU) 2023/988 [16] are still pending. Nevertheless, pursuant to the high-risk classification under the AI Act, businesses are required to adopt extensive documentation, monitoring, and risk mitigation procedures in the development of AI systems. These procedures include rigorous testing aimed at identifying and eliminating potential biases, maintaining transparency logs for decision traceability, and implementing comprehensive quality assurance throughout the lifecycle of the vehicle. Additionally, dedicated teams, including external auditors, are essential to conduct risk assessments and ensure ongoing compliance, rendering the Act's requirements technically complex, costly, and resource-intensive.

Although forthcoming sector-specific updates may clarify compliance obligations for automotive stakeholders, the alignment of these regulations with the AI Act remains uncertain, potentially impacting the entire supply chain. Automotive manufacturers are unable to promptly implement the stipulations of the AI Act in their current form. They may encounter new compliance obligations related to the updated Type Approval Framework standards as those standards are released. Balancing the specific requirements of the revised Type Approval Framework for regulated products with the overarching requirements of the AI Act will increase the complexity of AI compliance. Suppliers face a similar challenge.

The AI Act broadly applies to artificial intelligence systems introduced or utilised within the European Union (EU) or the European Economic Area (EEA), encompassing general-purpose models, irrespective of the geographic location of the supplier. This scope extends to operators situated outside the EU/EEA whose AI-derived outputs, such as predictions or recommendations, are employed within these regions. Consequently, non-EU companies, including manufacturers of autonomous vehicles—such as those headquartered in the United States, China, South Korea, or Japan—are mandated to comply with the AI Act if they introduce AI systems or utilise AI-generated results in the EU market. This extraterritorial jurisdiction necessitates that any automobile manufacturer or service provider entering the EU market fulfil overlapping regulatory obligations. For instance, a company based in the United States may be compelled to adhere to the EU's rigorous requirements for high-risk AI while simultaneously conforming to distinct or more lenient regulations in its home jurisdiction. The same provision applies, for example, to Chinese manufacturers entering the EU market with products primarily designed to satisfy their domestic standards. Numerous companies may adopt high EU standards on a global scale, potentially leading to a gradual harmonisation of AI security standards internationally, should other significant markets align with EU regulations. While this fosters regulatory consistency, it simultaneously escalates development costs and timelines.

4.5 Conclusions

Digital mobility services are being implemented in connection with the development of business models conceived through collaboration between vehicle manufacturers and technology firms. Digitalisation serves as a vital component in advancing these business models. It drives a transformation in consumer preferences as mobility services increasingly shift to online platforms through dedicated applications. Digitalisation facilitates the development of research and commercial applications for autonomous vehicles, whether they operate independently or in cooperative teams comprising both semi-autonomous and fully autonomous units. The prospect of employing autonomous vehicles alongside digital mobility services presents opportunities to customise a diverse array of these services to meet market demands. The extensive potential for digital mobility services is grounded in the 'X-as-a-service' concept elucidated in the existing literature. Prior to the deployment of autonomous vehicles, urban areas have emerged as central hubs for the proliferation of new mobility services, including ride-sharing platforms, which are expected to evolve alongside autonomous vehicles, supported by the establishment of business models centred on mobility services.

The term 'digital mobility service' has yet to be legally defined. Its meaning has been interpreted from the digital mobility business models that have been outlined. Reference to these models allows for the formulation of five distinct types of this service. Consequently, these services encompass the following provisions: (1) access to transport services offered by a third party utilising its vehicle, (2) access to a vehicle for the purpose of travel, (3) assistance to the driver in tasks directly associated with planning and executing the journey, (4) access to content (including voice and data) that, while not essential for planning and executing the journey, enhances the overall travel experience, and (5) provision of support for managing passenger and freight transport. These services are delivered via the internet by providers operating within an ecosystem that often includes the vehicle manufacturer. Moreover, these services are rendered for remuneration, functioning as voluntary services related to autonomous wheeled transport.

There is a notable deficiency in dedicated legal regulations concerning the provision of digital mobility services. Identifying appropriate legal sources should be conducted with careful consideration of the subject matter relating to these services, the variety of recipients who utilise them, and the modalities of their provision. The subject matter of digital mobility services should be delineated according to the various types of services recognised. This subject matter is examined from the perspective of the legal relationship established between the service provider and the recipient, which typically involves an operator of digital platforms. The recipients of digital mobility services encompass both consumers and business operators who incorporate these services into their operations. Consumers possess specific rights that are safeguarded under both EU law and the national legislation of Member

States. The manner in which digital mobility services are provided holds legal significance for several reasons. Firstly, regarding modes of wheeled transport, including the use of autonomous vehicles and artificial intelligence. Secondly, concerning the involvement of digital platforms in mediating the delivery of digital mobility services.

References

1. Act of 12 July 2024. Electronic Communications Law (Journal of Laws, item 1221)
2. Act of 15 November 1984 Transport Law (i.e. Journal of Laws 2024, item 1262)
3. Act of 18 July 2002 on provision of services by electronic means (i.e. Journal of Laws of 2024, item 1513, as amended)
4. Act of 21 August 1997 on real estate management (i.e. Journal of Laws of 2024, item 1145, as amended)
5. Act of 23 April 1964 Civil Code (i.e. Journal of Laws 2024, item 1061, as amended)
6. Aoyama, Y., Leon, L.F.A.: Urban governance and autonomous vehicles. Cities **119**, 103410 (2021)
7. Athanasopoulou, A., de Reuver, M., Nikou, S., Bouwman, H.: What technology enabled services impact business models in the automotive industry? An exploratory study. Futures **109**, 73–83 (2019)
8. BEREC 2021, BEREC Report on the interplay between the EECC and the EC's proposal for a Digital Markets Act concerning number-independent interpersonal communication services, 10 June 2021, BoR (21) 85, BEREC Report on the interplay between the EECC and the EC's proposal for a Digital Markets Act concerning number-independent interpersonal communication services. Accessed 21 Jan 2025
9. BEREC Report on OTT services, January 2016 for further information on this matter. https://berec.europa.eu/eng/document_register/subject_matter/berec/reports/5751-berec-report-on-ott-services. Accessed 21 Feb 2025
10. Bergman, N., Schwanen, T., Sovacool, B.K.: Imagined people, behaviour and future mobility: insights from visions of electric vehicles and car clubs in the United Kingdom. Transp. Policy **59** (2017)
11. Bissell, D.: Automation interrupted: how autonomous vehicle accidents transform the material politics of automation. Polit. Geogr.Geogr. **65**, 57–66 (2018). https://doi.org/10.1016/j.polgeo.2018.05.003
12. Bratton, B.H.: The Stack: On Software and Sovereignty. MIT Press, Cambridge, MA (2016)
13. Brown, B.: The social life of autonomous cars. Computer **50**(2), 92–96 (2017). https://doi.org/10.1109/MC.2017.59
14. Bustinza, O.F., Gomes, E., Vendrell-Herrero, F., Tarba, S.Y.: An organisational change framework for digital servitization: evidence from the Veneto region. Strateg. Chang.. Chang. **27**(2), 111–119 (2018). https://doi.org/10.1002/jsc.2186
15. See, for example: Judgment of the Court of 27 September 1988, Belgian State v René Humbel and Marie-Thérèse Edel, Case 263/86, European Court Reports 1988 -05365, ECLI:EU:C:1988:451; Judgment of the Court of 12 September 2000, Pavel Pavlov and Others v Stichting Pensioenfonds Medische Specialisten, Joined cases C-180/98 to C-184/98, ECLI:EU:C:2000:428; Judgment of the Court (Seventh Chamber), 11 September 2014, Sotiris Papasavvas v O Fileleftheros Dimosia Etaireia Ltd and Others, C-291/13, ECLI:EU:C:2014:2209
16. Cabanelas, P., Parkhurst, G., Thomopoulos, N., Lampon, J.F.: A dynamic capability evaluation of emerging business models for new mobility. Res. Transp. Bus. Manag.Manag. **47**, 100964 (2023). https://doi.org/10.1016/j.rtbm.2023.100964
17. Caprotti, F., Chang, I.C., Joss, S.: Beyond the smart city: a typology of platform urbanism. Urban. Transform. **4**(1), 4 (2022). https://doi.org/10.1186/s42854-022-00033-9. Epub Mar 26

18. Chawla, K.: An Overview of Relevant European Legislation for MSMEs Engaged in Data Sharing Centre of Excellence for Data Sharing & Cloud (coe-dsc.nl) (2024). https://resolver.tno.nl/uuid:55acffb9-1a40-44f7-aaa2-0a6d85dc8aa6. Accessed 21 Feb 2025
19. Consolidated version of the Treaty on the Functioning of the European Union (TFEU) [2012] OJ C326/26
20. Cunningham, M.L., Regan, M.A., Horberry, T., Weeratunga, K., Dixit, V.: Public opinion about automated vehicles in Australia: results from a large-scale national survey. Transp. Res. Policy Pract. **129**, 1–18 (2019)
21. Cusumano, M.A.: Staying Power: Six Enduring Principles for Managing Strategy and Innovation in an Uncertain World (Lessons from Microsoft, Apple, Intel, Google, Toyota and More). Oxford University Press, Oxford (2010)
22. de Vasconcelos, L., Farago, F., Facin, A., Flechas, X., Silva, L.: From open business model to ecosystem business model: a processes view. Technol. Forecast. Soc. Change **194**, Article 122668. https://doi.org/10.1016/j.techfore.2023.122668
23. De Vaujany, F.-X., Leclercq-Vandelannoitte, A., Holt, R.: Communities versus platforms: the paradox in the body of the collaborative economy. J. Manag. Inq. **29**, 450–467 (2019)
24. Directive (EU) 2018/1972 of the European Parliament and of the Council of 11 December 2018 establishing the European Electronic Communications Code (Recast) (Text with EEA relevance), PE/52/2018/REV/1, OJ L 321, 17.12.2018, pp. 36–214
25. Directive (EU) 2019/770 of the European Parliament and of the Council of 20 May 2019 on certain aspects concerning contracts for the supply of digital content and digital services (Text with EEA relevance), PE/26/2019/REV/1, OJ L 136, 22.5.2019, pp. 1–27
26. Directive (EU) 2019/771 of the European Parliament and of the Council of 20 May 2019 on certain aspects concerning contracts for the sale of goods, amending Regulation (EU) 2017/2394 and Directive 2009/22/EC, and repealing Directive 1999/44/EC (Text with EEA relevance), PE/27/2019/REV/1, OJ L 136, 22.5.2019, pp. 28–50
27. Directive 2011/83/EU of the European Parliament and of the Council of 25 October 2011 on consumer rights, amending Council Directive 93/13/EEC and Directive 1999/44/EC of the European Parliament and the Council and repealing Council Directive 85/577/EEC and Directive 97/7/EC of the European Parliament and of the Council Text with EEA relevance, OJ L 304, 22.11.2011, pp. 64–88
28. FLIX 2025. https://www.flixbus.co.uk/terms-and-conditions-of-carriage?_sp=5e0c3393-fc05-4f3c-a26c-c24b422d4bae.1743233612028&atb_pdid=276eb58e-04cd-4ce5-99c9-8ea770ea2d7a&_spnuid=20d2548a-0332-4d08-8799-aaa1e3ead59a_1743233611922. Accessed 12 Feb 2025
29. Ganco, M., Kapoor, R., Lee, G.K.: From rugged landscapes to rugged ecosystems: structure of interdependencies and firms' innovative search. Acad. Manag. Rev.Manag. Rev. **45**(3), 646–674 (2020). https://doi.org/10.5465/amr.2017.0549
30. Ganesh, M.I.: Entanglement. Peer-Rev. J. About **6**(1), 76–87 (2017). https://doi.org/10.7146/aprja.v6i1.116013
31. Gawer, A., Phillips, N.: Institutional work as logics shift: the case of Intel's transformation to platform leader. Organ. Stud. **34**, 1035–1071 (2013)
32. Geels, F.W., Schot, J.: Typology of sociotechnical transition pathways. Res. Policy **36**, 399–411 (2007)
33. Gekker, A., Hind, S.: Infrastructural surveillance. New Media Soc. **22**(8), 1414–1436 (2019). https://doi.org/10.1177/1461444819879426
34. Genzlinger, F., Zejnilovic, L., Bustinza, O.F.: Servitization in the automotive industry: how car manufacturers become mobility service providers. Strateg. Chang.. Chang. **29**(2), 215–226 (2020)
35. Hind, S., Kanderske, M., van der Vlist, F.: Making the car "platform ready": how big tech is driving the platformization of automobility. Soc. Media + Soc. **8**(2) (2022). https://doi.org/10.1177/20563051221098697
36. Hodson, M., McMeekin, A., Lockhart, A.: Urban infrastructure reconfiguration and digital platforms: who is in control? Environ. Innov. Soc. Trans.Innov. Soc. Trans. **50**, 100816 (2024)

37. Honda 2019: MONET Forms Capital and Business Partnership with Hino Motors and Honda. https://global.honda/en/newsroom/news/2019/c190328beng.html. Accessed 23 Feb 2025
38. Honda 2025, Honda Mobile Power Pack. https://global.honda/en/tech/Honda_Mobile_Power_Pack_MPP/. Accessed 23 Feb 2025
39. Humbs, C., Weller, M.: Is Tesla Violating Privacy Rules? Tagesschau (2020). https://web.archive.org/web/20200917225533/https://www.tagesschau.de/investigativ/kontraste/tesla-datenschutz-101.html. Accessed 21 Feb 2025
40. Iglezakis, I. (ed.): Legal Issues of Mobile Apps, a Practical Guide, p. 124. WoltersKluwer, Netherlands
41. Jacobides, M.G., Cennamo, C., Gawer, A.: Towards a theory of ecosystems. Strateg. Manag. J.. Manag. J. **39**(8), 2255–2276 (2018). https://doi.org/10.1002/smj.2904
42. James, A.: Honda announces Cruise AV testing program in Japan, ADAS & Autonomous Vehicle International, 9 Sept 2021. https://www.autonomousvehicleinternational.com/news/mobility-solutions/honda-announces-cruise-av-testing-program-in-japan.html. Accessed 23 Feb 2025
43. Judgment of the Court (Grand Chamber), 3 July 2012. UsedSoft GmbH v Oracle International Corp., Case C 128/11, ECLI:EU:C:2012:407
44. Judgment of the Court of Appeal in Gdańsk of 31.05.2016, II AKa 387/15, LEX no. 2200345
45. Judgment of the European Court of Justice of 16.03.2006, C-3/04, POSEIDON CHARTERING BV v. MARIANNE ZEESCHIP VOF, ALBERT MOOIJ, SJOERDTJE SIJSWERDA, GERRIT SCHRAM, ECLI:EU:C:2006:176
46. Kaminksi, M.: The Developing Law of AI Regulation: A Turn to Risk Regulation (Lawfare, April 21, 2022). https://www.lawfaremedia.org/article/the-developing-law-of-ai-regulation-a-turn-to-risk-regulation. Accessed 21 Apr 2024
47. Kapoor, R.: Ecosystems: broadening the locus of value creation. J. Organ. Des. **7**(1), 1–16 (2018). https://doi.org/10.1186/s41469-018-0035-4
48. Kenney, M., Zysman, J.: The rise of the platform economy. Issues Sci. Technol. **32**, 61–69 (2016)
49. Kohtämaki, M., Parida, V., Oghazi, P., Gebauer, H., Baines, T.S.: Digital servitization business models in ecosystems: a theory of the firm. J. Bus. Res. **104**, 380–392 (2019)
50. Krafft, M., Kumar, V., Harmeling, C., Singh, S., Zhu, T., Chen, J., Duncan, T., Fortin, W., Rosa, E.: Insight is power: understanding the terms of the consumer-firm data exchange. J. Retail. **97**(1), 133–149 (2021). https://doi.org/10.1016/j.jretai.2020.11.001
51. Lampón, J.F., Muñoz-Dueñas, P.: Are sustainable mobility firms reshaping the traditional relationships in the automotive industry value chain? J. Clean. Prod. **413**, Article 137522 (2023). https://doi.org/10.1016/j.jclepro.2023.137522
52. Leminen, S., Rajahonka, M., Wendelin, R., Westerlund, M., Nyström, A.: Autonomous vehicle solutions and their digital servitization business models. Technol. Forecast. Soc. Change **185**, Article 122070 (2022). https://doi.org/10.1016/j.techfore.2022.122070
53. Lippoldt, D.C., Strzyszowski, P.: Innovation in the Software Sector. OECD Publishing (2009). https://www.oecd.org/en/publications/innovation-in-the-software-sector_9789264076761-en.html. Accessed 12 Feb 2025
54. Mandal, A.K., Panarotto, F., Cortesi, A., Ferrara, F., Spoto, F.: Static analysis of Android Auto infotainment and on-board diagnostics II apps. Softw. Pract. Ex. **49**(7), 1131–1161 (2019)
55. Meyers, G., Van Hoyweghen, I.: "Happy failures": experimentation with behaviour-based personalization in car insurance. Big Data Soc. **7**(1), 1–14 (2020). https://doi.org/10.1177/2053951720914650
56. Monios, J., Bergqvist, R.: The transport geography of electric and autonomous vehicles in road freight networks. J. Transp. Geogr.Geogr. **80**, 102500 (2019)
57. Münzel, K., et al.: Explaining carsharing supply across Western European cities. Int. J. Sustain. Transp. 1–12 (2019)
58. Negoro, T., Ajiro, S.: An outlook of platform theory research in business studies. Waseda Bus. Econ. Stud. **48**, 1–29 (2012)

59. Nie, F., Li, X., Zhou, Ch.: Impact of privacy regulation involving information collection on the ride-hailing market. Omega **126**, 103077 (2024). https://doi.org/10.1016/j.omega.2024.103077
60. Parker, G.G., Alstyne, V., Marshall, W.: Two-sided network effects: a theory of information product design. Manag. Sci.. Sci. **51**, 1494–1504 (2005)
61. Perez-Moure, H., Lampon, J.F., Velano-Rodriguez, M.E., Rodríguez-Comesana, L.: Revolutionizing the road: how sustainable, autonomous, and connected vehicles are changing digital mobility business models. Eur. Res. Manag. Bus. Econ.Manag. Bus. Econ. **29**, 100230 (2023). https://doi.org/10.1016/j.iedeen.2023.100230
62. Regulation (EU) 2016/679 of the European Parliament and of the Council of 27 April 2016 on the protection of natural persons with regard to the processing of personal data and on the free movement of such data, and repealing Directive 95/46/EC, OJ L 119, 4.5.2016, pp. 1–88
63. Regulation (EU) 2022/1925 of the European Parliament and of the Council of 14 September 2022 on contestable and fair markets in the digital sector and amending Directives (EU) 2019/1937 and (EU) 2020/1828 (Digital Markets Act) (Text with EEA relevance), PE/17/2022/REV/1, OJ L 265, 12.10.2022, pp. 1–66
64. Regulation (EU) 2022/2065 of the European Parliament and of the Council of 19 October 2022 on a Single Market For Digital Services and amending Directive 2000/31/EC (Digital Services Act) (Text with EEA relevance), PE/30/2022/REV/1, OJ L 277, 27.10.2022, pp. 1–102
65. Regulation (EU) 2022/868 of the European Parliament and of the Council of 30 May 2022 on European data governance and amending Regulation (EU) 2018/1724 (Data Governance Act) (Text with EEA relevance), PE/85/2021/REV/1, OJ L 152, 3.6.2022, pp. 1–44
66. Regulation (EU) 2023/2854 of the European Parliament and of the Council of 13 December 2023 on harmonized rules on fair access to and use of data and amending Regulation (EU) 2017/2394 and Directive (EU) 2020/1828 (Data Act) (Text with EEA relevance), PE/49/2023/REV/1, OJ L, 2023/2854, 22.12.2023
67. Regulation (EU) 2023/988 of the European Parliament and of the Council of 10 May 2023 on general product safety, amending Regulation (EU) No 1025/2012 of the European Parliament and of the Council and Directive (EU) 2020/1828 of the European Parliament and the Council, and repealing Directive 2001/95/EC of the European Parliament and of the Council and Council Directive 87/357/EEC (Text with EEA relevance), PE/79/2022/REV/1, OJ L 135, 23.5.2023, pp. 1–51
68. Regulation (EU) 2024/1689 of the European Parliament and of the Council of 13 June 2024 laying down harmonised rules on artificial intelligence and amending Regulations (EC) No 300/2008, (EU) No 167/2013, (EU) No 168/2013, (EU) 2018/858, (EU) 2018/1139 and (EU) 2019/2144 and Directives 2014/90/EU, (EU) 2016/797 and (EU) 2020/1828 (Artificial Intelligence Act) (Text with EEA relevance), E/24/2024/REV/1, OJ L, 2024/1689, 12.7.2024, ELI: http://data.europa.eu/eli/reg/2024/1689/oj. Accessed 21 Feb 2025
69. Regulation (EU) No 181/2011 of the European Parliament and of the Council of 16 February 2011 concerning the rights of passengers in bus and coach transport and amending Regulation (EC) No 2006/2004 Text with EEA relevance, OJ L 55, 28.2.2011, pp. 1–12
70. Rietveld, J., Schilling, A.: Platform competition: a systematic and interdisciplinary review of the literature. J. Manag. 47, 1528–1529 (2021)
71. Rochet, J.-C., Tirole, J.: Platform competition in two-sided markets. J. Eur. Econ. Assoc. **1**(4), 990–1029 (2003)
72. Rode, P., Floater, G., Thomopoulos, N., Docherty, J., et al.: Accessibility in cities: transport and urban form. Disrupting mobility, pp. 239–273 (2017)
73. Ross, D.: Millennials Don't Care About Owning Cars, And Car Makers Can't Figure Out Why." Fast Company (2014). https://www.fastcompany.com/3027876/millennials-dont-care-about-owning-cars-and-car-makers-cant-figure-out-why. Accessed 21 Feb 2025
74. Sadowski, J.: When data is capital: datafication, accumulation, and extraction. Big Data Soc. **6**(1), 1–12 (2019). https://doi.org/10.1177/2053951718820549
75. SAE International: Taxonomy and Definitions for Terms Related to Driving Automation Systems for On-Road Motor Vehicles J3016_202104. https://www.sae.org/standards/content/j3016_202104/. Accessed 21 Feb 2025

76. Sarasini, S., Linder, M.: Integrating a business model perspective into transition theory: the example of new mobility services. Environ. Innov. Soc. Trans.Innov. Soc. Trans. **27**, 16–31 (2018)
77. Schüßler, E., Attwood-Charles, W., Kirchner, S., Schor, J.B.: Between mutuality, autonomy and domination: rethinking digital platforms as contested relational structures. Socioecon. Rev. **19**(4), 1217–1243 (2021). https://doi.org/10.1093/ser/mwab038
78. Sim, A.X.A., Sitohang, B.: OBD-II standard car engine diagnostic software development. In: 2014 International Conference on Data and Software Engineering, ICODSE), pp. 1–5 (2014)
79. Srnicek, N.: Platform capitalism. Polity Press, Cambridge, UK (2016)
80. Srnicek, N.: The challenges of platform capitalism: understanding the logic of a new business model. Juncture **23**, 254–257 (2017)
81. Steinberg, M.: From automobile capitalism to platform capitalism: toyotism as a prehistory of digital platforms. Eur. Group Organ. Stud. (2021). https://doi.org/10.1177/017084062110 30681
82. Stellantis: FCA and Waymo Further Expand Autonomous Driving Technology Partnership and Sign Exclusive Agreement for Light Commercial Vehicles (2020). https://www.media.stellantis.com/em-en/fca-archive/press/fca-and-waymo-further-expand-autonomous-driving-technology-partnership-and-sign-exclusive-agreement-for-light-commercial-veh icles. Accessed 9 Mar 2025
83. Stellantis: Stellantis Intensifies Electrification While Targeting Sustainable Double-Digit Adjusted Operating Income Margins in the Mid-term (2021). https://www.stellantis.com/en/news/press-releases/2021/july/stellantis-intensifies-electrification-while-targeting-sustai nable-double-digit-adjusted-operating-income-margins-in-the-mid-term. Accessed 7 Mar 2025
84. Stellantis: Fiat Celebrates the Success of New 500's 'Kiri' Project with 7,300 Users Generating over £635,000 Worth of Rewards (2022). https://www.media.stellantis.com/uk-en/fiat/press/fiat-celebrates-the-success-of-new-500-s-kiri-project-with-7-300-users-generating-over-635-000-worth-of-rewards. Accessed 7 Mar 2025
85. Stellantis: Intelligent Vehicles (2022). https://www.stellantis.com/en/technology/intelligent-vehicles. Accessed 14 Mar 2025
86. Stellantis: Amazon and Stellantis Collaborate to Introduce Customer-Centric Connected Experiences Across Millions of Vehicles, Helping Accelerate Stellantis' Software Transformation (2022)
87. Stilgoe, J.: Machine learning, social learning and the governance of self-driving cars. Soc. Stud. Sci. **48**(1), 25–56 (2017). https://doi.org/10.1177/0306312717741687
88. Stilgoe, J.: Seeing like a Tesla: how can we anticipate self driving worlds? Glocalism **3**, 1–20 (2017). https://doi.org/10.12893/gjcpi.2017.3.2
89. Sumaila, F., Bahsi, H.: Digital forensic analysis of mobile automotive maintenance applications. Forens. Sci. Int.: Digit. Invest. **43**(Suppl.), 301440 (2022). https://doi.org/10.1016/j.fsidi.2022.301440
90. Svennevik, E.M.C., Dijk, M., Arnfalk, P.: How do new mobility practices emerge? A comparative analysis of car-sharing in cities in Norway, Sweden, and the Netherlands. Energy Res. Soc. Sci. **82**, Article 102305 (2021). https://doi.org/10.1016/j.erss.2021.102305
91. Toyota 2019, Aioi Nissay Dowa Insurance and Toyota Develop Telematics Damage Service System, an Accident Response Service that Uses Connected Car Data, 22.12.2019. https://global.toyota/en/newsroom/corporate/31088545.html. Accessed 23 Feb 2025
92. Toyota 2020. E-palette. https://www.toyota-global.com/pages/contents/innovation/intell igent_transport_systems/world_congress/2018copenhagen/pdf/e-Palette_CONCEPT.pdf. Accessed 23 Feb 2025
93. Toyota 2021, Woven Capital Invests in Ridecell to Accelerate Global Growth in IoT-driven Automation for Mobility and Fleet Businesses, 17 Jan 2021. https://global.toyota/en/new sroom/corporate/35548927.html. Accessed 23 Feb 2025
94. Turienzo, J., Cabanelas, P., Lampon, J.F.: Business models in times of disruption: the connected and autonomous vehicles (uncertain) domino effect. J. Bus. Res. **156**, 113481 (2023). https://doi.org/10.1016/j.jbusres.2022.113481

95. UBER 2024. https://www.uber.com/legal/pl/document/?country=poland&lang=pl&name=general-terms-of-use. Accessed 21 Feb 2025
96. Venkatesh, R., Mathew, L., Singhal, T.K.: Imperatives of business models and digital transformation for digital services providers. Int. J. Bus. Data Commun. Netw. **15**, 105–124 (2019). https://doi.org/10.4018/ijbdcn.2019010107
97. Weber, H.: Pony, Investor Toyota Partner to 'Mass Produce' Robotaxis in China, 4 Aug 2023. https://techcrunch.com/2023/08/04/pony-investor-toyota-partner-to-mass-produce-robotaxis-in-china/. Accessed 23 Feb 2025
98. Zhao, J., Liang, B., Chen, Q.: The key technology toward the self-driving car. Int. J. Intell. Unmanned Syst. **6**(1), 2–20 (2018)
99. ZIP-ZAP Navigation. https://www.zap-map.com/app/route-plan. Accessed 23 Feb 2025
100. Jittrapirom, P., Caiati, V., Feneri, A.-M., Ebrahimigharehbaghi, S., González, M. J. A., Narayan, J.: Mobility as a service: A critical review of definitions, assessments of schemes, and key chal lenges. Urban Plan. **2**(2), 13 (2017) https://doi.org/10.17645/up.v2i2.931
101. Marres, N.: Co-existence or displacement: Do street tri als of intelligent vehicles test society? Br. J. Soc. **71**(3), 537–555 (2020) https://doi.org/10.1111/1468%204446.12730
102. Wigley E., Rose G.: Who's behind the wheel? Visioning the future users and urban contexts of connected and autonomous vehicle technologies. Geografiska Annaler: Series B, 2 April, pp. 1–37 (2020), https://doi.org/10.1080/04353684.2020.1747943

Chapter 5
The Significance of Digital Platforms in the Provision of Digital Mobility Services

5.1 Introduction

The current body of literature emphasises that digital mobility services, along with their associated business models, pertain to specific ecosystems that encompass various stakeholders [41, 53]. Hasselwander [32] asserts that, more recently, mobility platforms such as the 'unicorns', namely Uber, Bolt, Grab, GoJek, Didi Chuxing, and Careem, strive to amalgamate a diverse array of services, thereby rendering them increasingly multifaceted and convenient for users. Initially concentrating on ridesharing with superior matchmaking capabilities compared to traditional taxis, these platforms now underscore the value proposition of multifunctionality, with ridesharing constituting merely one of many features.

Additionally, Eisenmann et al. [22] introduced a triangular framework pertaining to legal relationships in platform dynamics. By elucidating this relationship, they posited that users engage in transactions with one another while simultaneously establishing affiliations with platform providers.

For instance, video game networks consist of two distinct user groups: players and developers. Developers engage in the sale of games to players, representing the initial set of exchanges within a video game network. Additionally, developers are required to enter into contracts with the platform's provider (e.g., Nintendo) to obtain permission for game publication, constituting the second set of exchanges. Lastly, players are necessitated to acquire a console from the platform provider, which forms the third set of exchanges. Consequently, three types of relationships may be established. These relationships can be articulated in accordance with the example provided below.

Platforms are classified as two-sided when they accommodate two distinct groups of users that mutually attract one another, such as video game players and developers [39, 40, 48]. Typically, two-sided networks comprise a supply side made up of sellers providing complements to users on the demand side. Users on one side generally maintain consistent transactional roles rather than interchanging them. In

A. Krasuski, *Legal Challenges of a Transformation from a Product-Centric to a Service-Centric Business Model and Procedures*, Studies in Systems, Decision and Control 610, https://doi.org/10.1007/978-3-031-98901-8_5

contrast, one-sided networks consist of users who share identical roles, public telephone networks serve as a prime example, wherein users are categorised as either callers or receivers.

Every network mediated by a platform possesses a central hub, while other platforms may assume subordinate roles within the network as supply-side users or component providers. The network may be governed by a proprietary platform, which suggests a singular company as the exclusive supplier (e.g., Nintendo's Wii), or it may incorporate multiple suppliers that offer competing yet compatible versions of a standard platform (e.g., Ubuntu Linux vs. Red Hat Linux). Users who transition between competing providers of a common platform generally retain their investments in platform-specific add-ons or their understanding of platform regulations [21]. Platform markets consist of sets of competing platforms, each enabling distinct networks (for instance, the video game market includes Xbox, Playstation, and Wii platforms).

Platform markets typically exhibit a limited number of competing platforms; frequently, a significant portion of users is drawn towards a single platform (e.g., Microsoft's Windows, Adobe's PDF, eBay's online auctions). The presence of platforms within a market is generally constrained when network effects are pronounced, when individual users experience considerable costs associated with multi-homing (establishing affiliation with multiple platforms), and when there exists minimal user demand for distinct platform functionalities [4, 23, 37, 54].

Considering that digital mobility business models necessitate collaboration among various stakeholders, it is essential to differentiate between two categories of these stakeholders. The distinction between the first group and the second hinges on whether the legal relationship with the entity is supported by civil law provisions or governed by administrative law. These two concepts require further clarification.

The first category includes private entrepreneurs who collaborate in the implementation of a business model centred on providing digital mobility services. These participants create value for customers by offering digital mobility services, such as car-sharing firms and digital platform service providers, along with the technologies that support such services, including vehicle connectivity software and positioning and localisation technologies [41].

Even mobility services are typically executed by third-party companies, with oversight provided by the manufacturer. This necessitates the consideration of two distinct value chains, wherein dealers and mobility service providers encompass the majority of the respective chain [30].

Entities comprising the initial group engage in legal relationships. The Polish Civil Code [50] governs civil law relations between specific entities, as stipulated in Article 1. A civil law relationship is characterised as a legal relationship (i.e., a social relationship regulated by law) defined by the provisions of civil law norms. Nevertheless, the very notion of "legal relationship" provokes ambiguities within the literature [2]. A civil law relationship may manifest in both pecuniary and non-pecuniary forms. An illustrative example of such a legal relationship is a contract. A civil law relationship emerges (and is also subject to modification or termination) as a consequence of particular legal events. These events include those cited in the

Polish Civil Code [50] that are independent of human volition (e.g., the passage of time, birth, death) as well as those contingent upon it, including legal actions, court judgments, and administrative acts that result in civil law effects. Civil law effects are also generated by torts and unjust enrichment [47].

Civil law relationships encompass natural persons, legal persons, and organisational units that, although not classified as legal persons, have been endowed with legal capacity by virtue of law. This capacity, referred to as legal subjectivity [31], is established through legal statutes. Every human being inherently possesses this status due to their intrinsic 'feature of humanity,' whereas for organisational units, their status is conferred based on the discretion of the legislator. Consequently, while every individual qualifies as a natural person, not all organisational units are designated as legal persons or possess legal capacity. Furthermore, the terms' subjects' and parties' in civil law relationships are frequently utilised interchangeably, however, they remain distinct in their meanings. Typically, only two parties are involved, although situations such as co-ownership or civil partnerships may involve additional parties. Moreover, each party may comprise multiple entities, exemplified by a number of lessees in a rental agreement or several injured parties (creditors) in a compensation context. The rights and obligations of the parties delineate the essence of a civil law relationship, whereby one party's entitlement correlates with the obligation of the other party. An entitlement arising from this relationship is designated as a subjective right.

This indicates that the utilisation of digital platforms necessitates the engagement of entities, including both natural persons and legal entities, within intricate civil law relationships. In relation to these legal relationships, substantial volumes of data are processed, which also encompass personal data as defined under Article 1(4) of Regulation (EU) 2016/679 [42].Consequently, it is essential to delineate the role of these entities in the processing of personal data while considering these complex civil legal relationships.

In relation to the armed conflict in Ukraine, European Union regulations were established in the form of EU Regulation 833/2014 [14], which imposes restrictions on cooperation with entrepreneurs from Russia concerning specific subject matters and objects. Notably, access to both primary and secondary capital markets within the EU has been restricted for certain Russian banks and companies. A prohibition on the direct or indirect purchase, sale, or provision of investment services, as well as assistance in the issuance of, and other activities involving, transferable securities and money market instruments, has been implemented with respect to, among others, the Government of Russia, the Central Bank of Russia, legal persons, entities, and bodies acting on their behalf, as well as entities listed in Annexes V and VI of Council Regulation (EU) 833/2014.

The second category encompasses public authorities, as they may influence the performance of services within their jurisdiction. According to Cabanelas et al. [3], public authorities, in particular, are regarded as a distinct case due to their potential, which is dependent on legal competencies, to regulate and facilitate or constrict the introduction of technological advancements. Even in instances where regulations are established centrally, their capacity to facilitate and incentivise can prove critical.

Specifically, they possess the ability to invest in appropriate infrastructure, promote networking and collaboration among and within industries, and adopt policy positions that prioritise the adoption of new technologies as essential to local economic or transport networks. In these manners, they can substantially contribute to the establishment of the mobility ecosystem.

The conduct of regulatory authorities may impose obstacles to the implementation of digital mobility services within the framework of the provision of such services.

In the examination of the relationship between an individual and the state (authority), the discipline of administrative law has established two fundamental concepts: the legal-administrative relationship and the legal-administrative situation. The former pertains to the objectivist perspective on the functioning of administrative law, whereas the latter is characterised by a subjective approach, particularly in the context of exercising sovereignty [19]. The legal-administrative situation of a particular entity is delineated exclusively by the provisions of substantive law, within which legal possibilities or obligations are acknowledged, as well as the extent of legally effective actions and the scope of accountability for contraventions of the law. The legal-administrative relationship is differentiated by the correspondence between the rights and obligations of distinct entities and those of other entities, including specific claims from both parties in this legal relationship concerning their adherence to or fulfilment of the established obligations [29].

5.2 Platforms as Gatekeepers and Very Large Platforms

Competition law offers a remedial framework when it can be demonstrated that an undertaking is engaging in the abuse of its dominant position, as outlined in Article 102 of the Treaty on the Functioning of the European Union (TFEU) [13]. Such abuse occurs when the market power of the undertaking is regarded as excessively formidable [35].

Motta and Peitz [36] rightly assert that competitive distortions can arise in specific markets under certain circumstances. The factors contributing to these distortions may include:

(1) The characteristics of the market that are not necessarily attributable to the behaviour of firms, including economies of scale or scope, network effects, switching costs and lock-in effects, and consumer behaviour; or
(2) The behaviour of the market participants themselves, such as, among other factors, tacit or explicit collusion, various horizontal or vertical agreements, contractual stipulations imposed on consumers, and other forms of abusive practices.

The two conditions delineated above may manifest concurrently within the same market. As noted by Laux et al. [35], the emergence of dominant platforms is attributable not exclusively to the actions of the firms but also to the inherent structure of the markets themselves.

Competition law reform, recognising the aforementioned market phenomena, has an international dimension. The evolution of competition law aims to ensure that it does not hinder digital and sustainable development within the realm of new technologies. Concerns regarding the adverse implications of the enforcement of this law [49] are driving the demand for a more comprehensive, expedited, and intrusive framework of competition law [5].

The Digital Markets Act [44] establishes a regulatory framework within the digital economy, designed to clarify the essential legal relationships between enforcement authorities of EU competition law and regulated enterprises [1]. Comeing [6] asserts that the European Union aims to reform the business practices of large technology firms through the implementation of the DMA. Enterprises that exploit their market power will be required to share the data they acquire with competitors and regulators, prohibited from granting themselves preferential access to adjacent markets, and restricted from pursuing mergers. The DMA seeks to prevent the abuse of dominant positions by these businesses, enhancing consumer choice and control over their data, and primarily aims to promote the growth of smaller businesses, ensuring they are not marginalised by the influence of more powerful competitors. In contrast to existing competition regulations, which have shown inefficacies in the digital era, the DMA, as a new legislative measure, will initiate preventive actions by imposing penalties for ex-post abuses of dominance.

The initial legislative framework consisted of a single regulation governing the digital platform market. The Digital Markets Act was subsequently distinguished from the Digital Services Act [45] to implement ex-ante rules specifically applicable to gatekeeper platforms. Furthermore, the DSA incorporates a specialised regime for very large online platforms ('VLOPS'). The following observations regarding the interpretation of the DSA provisions also relate to platforms classified as very large online search engines. In contrast, the concept of market size for 'gatekeepers-turned-VLOPS' now serves a different purpose, indicating the ex post social risk arising from the content disseminated through extensive platforms. This development may have adverse implications for the effectiveness of the oversight mechanisms stipulated by the DSA, particularly concerning independent audits and risk assessments [35].

The rules governing competition must be tailored to the unique circumstances under which new digital business models operate [24]. Given their systemic significance, certain online platforms, which serve as 'private gatekeepers of markets, customers, and information', may pose a threat to the equity and transparency of markets. Given that 'competition policy alone cannot address all the systemic problems that may arise in the platform economy', it may be necessary to establish supplementary regulations to ensure competition, fairness, innovation, and market entry. Specifically, the Commission has announced the initiation of a sector inquiry to evaluate the effectiveness of the existing competition rules and has indicated its intention to assess whether ex ante regulatory measures may be required to ensure a competitive market concerning gatekeeping platforms that exhibit substantial network effects [25]. Moreover, the Commission has initiated a public consultation regarding the necessity of a potential new competition tool that would effectively and timely address structural competition challenges by implementing behavioural and, where deemed appropriate, structural remedies [26]. In December 2020, the Commission ultimately unveiled its proposal for a Digital Markets Act [27]. Crémer et al. [15] underscore two principal and consistently emphasised concepts fundamental to the DMA, specifically 'fairness' and 'contestability'.

Moreover, the special regime established for very large online platforms seems to serve as a form of protection for emerging European competitors against certain obligations predominantly faced by U.S.-based platforms. If this is correct, the VLOP category is not relevant to the fundamental theory of harm outlined in the Digital Services Act, particularly regarding the management of 'systemic risks' as described in Article 34 of the DSA. The European Electronic Communications Code [18] includes a distinct regime for communications providers that hold 'significant market power' ('SMP'). However, it is important to note that it applies principles of competition law rather than societal risks as detailed in the DSA. See Article 64(2) of Directive (EU) 2018/1972.

Moreover, the specialised framework for very large online platforms appears to be designed to protect emerging European competitors from certain obligations that are mainly applicable to platforms based in the United States. If this proposition is valid, the classification of VLOPS is not inherently aligned with the primary theory of harm articulated in the Digital Services Act, particularly regarding the management of "systemic risk" as outlined in Article 34 of the DSA. The European Electronic Communications Code [18] includes a specialised framework for communication service providers that possess "significant market power" ('SMP'). However, it relates to concepts of competition law rather than societal risks, which are addressed under the DSA, especially in Article 64(2) of Directive (EU) 2018/1972 [35].

The Digital Markets Act and the Digital Services Act focus on aspects of digital platforms, particularly gatekeepers and VLOPS, albeit with differing objectives. The criteria established by the DMA for gatekeepers are based on the economic strength of the platforms, while the DSA purposefully outlines a separate category in which VLOPS operate as de facto public spaces, defined by user volume [28].

5.3 Platforms Governed by the Digital Markets Act

The Commission has taken measures to ensure that the core platform services ('CPS') are competitive and fair, regardless of the actual, potential, or presumed impacts resulting from the actions of a designated gatekeeper. The term 'gatekeeper' is explained in the Digital Markets Act under Article 2(1) as 'an undertaking providing core platform services, designated according to Article 3 of the DMA' [16].

To be formally classified as a gatekeeper under the Digital Markets Act ('DMA'), a digital platform service provider must meet three qualitative criteria specified in Article 3: (1) it possesses substantial influence on the internal market; (2) it offers a core platform service that acts as a critical gateway for business users to connect with end users; and (3) it maintains an entrenched and durable position in its operations, or it is likely to sustain such a position in the near future.

An initiative will be regarded as meeting the aforementioned requirements when, in accordance with Article 2(2) of the DMA, the following quantitative criteria are fulfilled, specifically concerning the Core Platform Service ('CPS'):

(1) It has achieved an annual turnover of at least EUR 7.5 billion in each of the last three financial years, or its average market capitalisation or equivalent fair market value amounted to no less than EUR 75 billion in the last financial year, and it provides the same core platform service in a minimum of three Member States.
(2) It has provided a core platform service that in the last financial year had no fewer than 45 million monthly active end users established or located within the Union and at least 10,000 yearly active business users established in the Union.
(3) The thresholds outlined in point 2 were met in each of the last three financial years. For any CPS provider that does not comply with these quantitative standards, the European Commission may designate it as a gatekeeper if the qualitative criteria are satisfied.

To date, the Commission has designated a total of 24 gatekeepers that provide core platform services [10]. On 6 September 2023, the European Commission designated, for the first time, six gatekeepers—Alphabet, Amazon, Apple, ByteDance, Meta, and Microsoft—under the Digital Markets Act. On 29 April 2024, the Commission designated Apple concerning its iPadOS, the operating system for tablets, as a gatekeeper under the DMA. Furthermore, on 13 May 2024, the Commission designated Booking.com as a gatekeeper for its online intermediation service, Booking.com, under the DMA. In total, 24 core platform services offered by these gatekeepers have been designated.

Article 2, point (2) of Regulation (EU) 2022/1925 enumerates ten categories of core platform services ('CPS'), specifically: (a) online intermediation services, (b) online search engines, (c) online social networking services, (d) video-sharing platform services, (e) number-independent interpersonal communications services, (f) operating systems, (g) web browsers, (h) virtual assistants, (i) cloud computing services, and (j) online advertising services, which includes any advertising networks, advertising exchanges, and any other advertising intermediation services provided by an undertaking that offers any of the CPS outlined in points (a) to (i). To ascertain whether a service offered by an undertaking qualifies as a CPS that fulfils the criteria established in Article 3(1)(b) of Regulation (EU) 2022/1925, it is imperative, as an initial step, to delineate the service in question. A comprehensive analysis predicated on various provisions of the Digital Markets Act must be conducted to accurately delineate such a service, which is inherently complex. It is necessary to identify potential distinct CPSs based on Section D, paragraph 2 of the Annex to Regulation (EU) 2022/1925. Furthermore, Article 13(1) of Regulation (EU) 2022/1925 stipulates that no action taken by an undertaking providing CPSs, which entails segmenting, dividing, subdividing, fragmenting, or splitting those services through contractual, commercial, technical, or any other means to elude the quantitative thresholds specified in Article 3(2) of that regulation, shall inhibit the Commission from designating the undertaking as a gatekeeper pursuant to Article 3(4) of the same regulation. As indicated in recital (11) of the preamble to the DMA, that regulation aims to achieve an objective that is complementary to, yet distinct from, that of EU competition rules, which is to safeguard undistorted competition in any given market. Therefore, the enforcement of EU competition rules, including established competition law precedents, does not preclude the application of the DMA, and vice versa. Consequently, the delineation of CPSS under the DMA does not affect the definition of the relevant market for the application of EU competition rules (and vice versa), leading to potentially divergent outcomes in these two types of analyses.

One category of these CPS encompasses online intermediation services. According to Article 5(2), the term 'online intermediation services' refers to the services as defined in Article 2, point (2), of Regulation (EU) 2019/1150 [43]. According to this definition, 'online intermediation services' are characterised by fulfilling all of the following criteria:

(a) they qualify as information society services as delineated in point (b) of Article 1(1) of Directive (EU) 2015/1535 [17];
(b) they enable business users to present goods or services to consumers, with the objective of facilitating the initiation of direct transactions between those business users and consumers, regardless of the location where those transactions are ultimately finalised;
(c) they are delivered to business users based on contractual agreements established between the service provider and the business users who offer goods or services to consumers.

The Digital Markets Act establishes a variety of obligations for gatekeepers, particularly those providing online intermediation services. Articles 5–7 of the DMA outline several obligations that significantly affect online intermediation services classified as CPS. Furthermore, these obligations explicitly confer enforceable rights to competing online intermediation services.

Under Article 5(3) of the DMA, gatekeepers are required to allow commercial users to distribute their products on various—preferably superior—terms through competing online intermediation services or via their own sales channels. This prohibition covers not only extensive parity clauses that impose the obligation not to offer more advantageous terms through alternative sales channels than those provided by the gatekeepers' online intermediation services, but also includes parity clauses compelling commercial users to propose terms that are at least as favourable through the gatekeepers' online intermediation services as through third-party online intermediation services. Furthermore, the prohibition includes narrow parity clauses, specifically the obligation not to offer differing or enhanced terms through their own sales channels.

The Digital Markets Act aims to enhance the differentiation between the terms under which products and services are offered, thereby improving price competition for the benefit of end users and fostering competition among rival online intermediation services provided by gatekeepers.

Commercial users of the 'gatekeepers' online intermediation services have the right to distribute their products and services through competing online intermediation services as well as their own sales channels without any restrictions. Moreover, similar to rival online intermediation services, they are encouraged to report violations by gatekeepers to the Commission, either by filing a complaint or using the newly established whistleblowing tool.

Furthermore, the prohibition of 'self-preference' as stipulated in section 6(5) of the DMA is of considerable significance. Gatekeepers are barred from prioritising their products in rankings, thereby facilitating easier access in comparison to similar competing products. The ranking process is determined algorithmically, dictating how goods and services are visually displayed or acoustically reproduced. The definition is intentionally broad; it suffices that only one result is presented and specifies the manner of presentation. Additionally, gatekeepers are mandated to ensure that the ranking is transparent, fair, and non-discriminatory.

Through this regulation, the Digital Markets Act aims to prevent gatekeepers from exploiting their substantial market position in core platform services ('CPS') markets to extend their market power to downstream services and products without facing competition based on merit. Furthermore, this regulation seeks to eliminate the preservation of existing market positions through the practice of self-privileging.

The self-privileging ban established by the Digital Markets Act is more comprehensive than the previously mentioned decision. It applies to all core platform services and specifically concerns online intermediation services, such as online marketplaces, application stores, price comparison websites, hotel and flight booking platforms, as well as search engines. Competing online brokerage services are entitled to fair classification and must not be disadvantaged compared to the gatekeeper's proprietary online brokerage services. These services are encouraged to report any infringements committed by gatekeepers to the Commission, either through a formal complaint or by utilising the newly established whistleblowing tool. Although the Commission serves as the sole enforcement authority of the DMA, with national authorities taking on a supplementary role, national authorities are permitted to conduct preliminary investigations if authorised by the relevant Member State, thereby providing a primary point of contact for third-party complaints.

In addition to the public enforcement of the Digital Markets Act by the Commission, online intermediation services possess the option to pursue private enforcement. In instances where gatekeepers violate obligations established under the DMA, affected parties are afforded the opportunity to initiate civil actions for injunctions, including interim relief, before the national courts within the EU and to file claims for damages as deemed appropriate. Subsequently, once the Commission has identified a breach of the DMA, follow-on actions for damages against gatekeepers will also become viable.

For instance, under the 11th Amendment to the German Act Against Restraints of Competition (Gesetz gegen Wettbewerbsbeschränkungen, GWB), the German legislature has expanded the binding effect of Commission decisions, which are well-known in antitrust law, to encompass Commission decisions regarding the Digital Markets Act [38].

The Commission has developed a Whistleblower Tool [11], which will enable parties to submit anonymous reports. In addition to the public enforcement of the Digital Markets Act by the Commission, online intermediation services are also afforded the opportunity to engage in private enforcement.

5.4 Platforms Governed by the Digital Services Act

To assess the applicability of the Digital Services Act to digital mobility services, it is essential to first scrutinise the extent to which the various categories of digital mobility services pertain to the DSA, while taking into account the specific categories of these services as delineated in Chap. 4.

Initially, these determinations must be predicated upon Article 2(1) in conjunction with Article 1(2) of the Digital Services Act. According to the first of the aforementioned provisions, the DSA is applicable to intermediary services offered to service recipients possessing a place of establishment or a physical presence within the European Union, irrespective of the location of the providers of such intermediary services. Thus, the criterion for the applicability of the DSA to digital mobility services entails ascertaining whether these services conform to the definition of an 'intermediary service'. The European Union legislator has delineated the concept of 'intermediary service' in Article 3(g) of the DSA, characterising it as an 'information society service', which is manifested in the three forms specified in that provision. Therefore, it is imperative to elucidate the meaning of the term 'information society service'. To clarify the term 'information society service', the EU legislator referenced Article 3(a) of the DSA in relation to Article 1(1), point (b), of Directive (EU) 2015/1535, thereby eliminating any ambiguity concerning the interpretation of this concept. This definition articulates that an information society service is any service typically rendered for remuneration, at a distance, via electronic means, and at the specific request of a service recipient. The legislator elucidates the components of the information society service in the following manner. The term 'at a distance' signifies that the service is delivered without the simultaneous presence of the parties involved. The phrase 'by electronic means' indicates that the service is initially sent and subsequently received at its designated location utilising electronic equipment for processing (including digital compression) and data storage, and is entirely transmitted, conveyed, and received through wire, radio, optical means, or other electromagnetic methodologies. The phrase 'at the individual request of a recipient of services' connotes that the service is provided through the data transmission requested by the individual. An indicative list of services excluded from this definition is outlined in Annex I of Directive (EU) 2015/1535.

Each category of digital mobility service displays the attributes of an information society service as defined in Article 1(1), point (b), of Directive (EU) 2015/1535. Consequently, the applicability of the Digital Services Act depends on the existence of at least one of the types of 'intermediary service' specified in Article 3(g)(i–iii) of the DSA regarding these services.

According to Article 3(g) of the DSA, an 'intermediary service' includes one of the following categories: a 'mere conduit' service, a 'caching' service, or a 'hosting' service.

In recital 29 of the DSA, the legislator elucidates the interpretation of the term 'intermediary service'. The legislator emphasises that 'intermediary services' encompass a broad spectrum of economic activities conducted online, which continually evolve to facilitate swift, safe, and secure information transmission, ensuring convenience for all participants in the online ecosystem. The 'mere conduit' service constitutes the transmission of information via a communication network, provided by a recipient of the service, as well as the provision of access to a communication network. For clarity, the term 'recipient' refers to any natural or legal person utilising an intermediary service, particularly for the purpose of seeking information or making it accessible (Article 3(b) of the DSA). As illustrations of 'mere conduit' services, the legislator enumerates the following in recital 29 of the DSA: internet exchange points, wireless access points, virtual private networks, DNS services and resolvers, top-level domain name registries, registrars, certificate authorities that issue digital certificates, voice over IP, and other interpersonal communication services.

The second category of information society services encompassed within 'intermediary services' pertains to a 'caching' service. This service involves the transmission of information via a communication network, as provided by the recipient of the service, and includes the automatic, intermediate, and temporary storage of such information. This storage is carried out solely to enhance the efficiency of the information's subsequent transmission to other recipients upon their request, as articulated in Article 3(g)(ii) of the Digital Services Act. Furthermore, according to recital 29 of the DSA, these services are essential for ensuring the seamless and effective transmission of information disseminated on the internet.

The third category of services outlined in Article 3(g)(iii) of the Digital Services Act, classified as 'intermediary services', includes hosting services. These are characterised by the storage of information supplied by, and at the request of, a recipient of the service. In recital 29 of the DSA, it is noted that such services encompass various categories, such as cloud computing, web hosting, paid referencing services, or services that facilitate the sharing of information and content online, including file storage and sharing. Intermediary services may be provided independently, as part of a different type of intermediary service, or concurrently with other intermediary services. The classification of a specific service as a 'mere conduit', 'caching', or 'hosting' service depends solely on its technical functionalities, which may evolve over time, and should be assessed on a case-by-case basis.

The term 'hosting service' represents an 'online platform', as delineated in Article 3(i) of the DSA. In accordance with this definition, an 'online platform' is classified as a hosting service. The scope of this service encompasses, at the request of a recipient of the service, the storage and dissemination of information to the public, unless that activity constitutes a minor and merely ancillary feature of another service or a minor functionality of the principal service. Furthermore, for objective and technical reasons, this ancillary functionality cannot be utilised independently of the other service, nor may the integration of such a feature or functionality into the other service serve as a means to circumvent the applicability of the DSA. To attain a comprehensive understanding of this specific type of hosting service, it is pertinent to reference the definition of 'dissemination to the public', as articulated in Article 3(k) of the DSA. This concept signifies the act of making information accessible, at the request of the service recipient who provided the information, to a potentially unlimited number of third parties.

In accordance with the aforementioned definitions under the DSA, one may draw the following conclusion regarding the applicability of types of the DSA to services that encompass digital mobility services, as delineated in Chap. 4. Generally speaking, the first and second categories of digital mobility services, which involve facilitating access to passenger and goods transportation services, as well as providing access to a mode of transport, do not fundamentally constitute an 'intermediation service' as defined in Article 3(g) of the DSA. Through the digital platform operator, basic information is relayed between individuals seeking to transport either a person or goods and prospective drivers, thereby forming the essentialia negotii of the contract for the transportation of goods or services. The range of information transmitted is predetermined by the digital platform operator. This designated scope of information is intended to enable available drivers to indicate their readiness to provide the transportation service. The DSA does not encompass within its framework the regulations governing the conclusion of the transportation contract or its execution. The factual activities outlined above do not fall under the category of 'mere conduit' service, as defined in Article 3(g)(i) of the DSA, since the provision by the service provider does not entail the offering of access to the communication network. Conversely, the provider of this service is obligated to relay the request for the use of the transportation service to drivers who are available and interested in undertaking the journey.

The diverse categories of services, including 'mere conduit', 'caching', and 'hosting', are articulated in a comparable manner within Directive 2000/31/EC (E-Commerce Directive). The Digital Services Act has adopted these definitions from the E-Commerce Directive. Although recital 29 of the DSA acknowledges wireless access points, virtual private networks (VPNs), DNS services, DNS resolvers, as well as services from top-level domain name registries, registrars, and certification authorities responsible for issuing digital certificates, it also includes, at least concerning the exemption from liability, Internet voice telephony (VoIP). Concurrently, it indicates that the scope of these services encompasses 'other interpersonal communication services'. In the Judgment of the Court (Third Chamber) dated 15 September 2016, C-484/14 [33], the Court of Justice of the European Union (CJEU) examined the issue of mere conduit services, specifically concerning the provision of Wi-Fi network services, asserting that "the provision of the service referred to in that article must involve the transmission in a communication network of information" (recital 46). Julià-Barceló and Koelman [34] contend that the 'mere conduit activity' incorporates both the role of transmitting information (for instance, providing cables and routers) and the facilitation of access to a communication network such as the Internet.

Truyens and Van Eecke [51] articulate that to benefit from the additional protection regime, 'the provider of the mere conduit services must not initiate the transmission, must not select the receiver, and must not select or modify the information that is transmitted'. With respect to the first and second categories of digital mobility services, the provider of these services transmits the request to a limited group of recipients, namely, drivers who are in close proximity to the individual requesting the ride. In this framework, within the two types of digital mobility services analysed, as described in Chap. 4, the 'mere conduit service' is not applicable due to the absence of a fundamental characteristic inherent to this service. Mere conduit services resemble traditional conveyance services; however, they constitute a minor aspect of a provider's comprehensive service offering [52]. Therefore, an entity offering digital mobility services of the first type would be excluded from the applicability of the DSA concerning the provision of such services, subject to the note below.

An entity providing the second category of digital mobility services may be subject to the Digital Services Act in various circumstances. Chapter 4 describes digital mobility services, illustrating an example in which club members exchange unique vehicles for remuneration, facilitated through an online digital portal. If the sequence of events were reversed—whereby, upon the recipient's request, an offer to rent unique vehicles is first presented online, followed by the placement of an order for a specific vehicle based on that offer—it becomes imperative to ascertain whether this service qualifies as a 'hosting' service, and additionally, whether it may also be classified as an 'online platform' service. In this regard, a thorough analysis of the individual elements defining an 'online platform' is essential.

In principle, a 'hosting' service, as delineated in Article 3(g)(iii) of the Digital Services Act, pertains to the storage of information that is provided by and at the behest of the recipient of the service. The 'recipient of the service' utilises an intermediary service to seek information or to render it accessible. Both objectives of the recipient would thus be accomplished. Concurrently, such a service necessitates a thorough examination of a more specific definition relevant to 'hosting' services, namely the definition of 'online platforms'. According to Article 3(i) of the DSA, the term 'online platform' encompasses a hosting service that, at the request of a recipient of the service, stores and disseminates information to the public.

The legislator defined the term 'dissemination to the public' in Article 3(k) of the DSA, which refers to the act of making information available, at the request of the service recipient who provided the information, to a potentially unlimited number of third parties. The EU legislator, in recital 14 of the preamble to the DSA, adopted a comprehensive interpretation of the term 'dissemination to the public', indicating that this term encompasses the provision of information to a potentially unlimited number of persons, thereby facilitating easy access to the information by service recipients in general, without any additional action required from the service recipient providing the information, regardless of whether those individuals actually access the information in question. Concurrently, the recital states that if access to the information necessitates registration or admission to a specific group of service recipients, such information should be regarded as disseminated to the public. In this instance, the EU legislator establishes the condition that service recipients desiring access to the information must be automatically registered or admitted without any human decision or selection regarding access. If, in the described scenario, this condition is satisfied, the 'online platform' service could be applied. Consequently, if the service is disseminated to an unlimited number of recipients, it would classify as an online platform service. In this context, the service under discussion could be likened to the offerings of the Booking.com platform. The EU Commission categorised these services as hosting services within the provisions of Article 3, point (g)(iii) of the DSA, which stores and disseminates information to the public at the request of its service recipients. Thus, it qualifies as an online platform according to Article 3, point (i), of the DSA [7].

The third, fourth, and fifth categories of digital mobility services, as delineated in Chap. 4, are made available through applications accessible in app stores. The functioning of these app stores should be recognised as an 'online platform' in accordance with Article 3(i) of the Digital Services Act. It is pertinent to reference the position of the European Commission [8] regarding the legal classification of the App Store (Google), as documented in the Commission decision dated 25 April 2023. Point 1 of the preamble to that decision stipulates: 'The App Store is classified as a hosting service according to Article 3, point (g)(iii), of Regulation (EU) 2022/2065, as it stores and disseminates information to the public at the request of recipients of its service. Consequently, it is categorised as an online platform pursuant to Article 3, point (i), of Regulation (EU) 2022/2065'. Moreover, the functions of the App Store are articulated in point 12 of the preamble to the aforementioned decision as follows: 'Firstly, the intended use of the App Store by service recipients, specifically developers and end-users, remains consistent, or largely analogous, regardless of the device on which the online platform is accessed. The App Store serves to host applications developed by developers at their request and distributes them to the public for potential usage by end users across various devices. Developers utilise the App Store to list, distribute, and update their applications. End users engage with the App Store to search for, download, or update applications, whether free or paid, and to procure digital content or services within the applications they have downloaded. Contrary to Apple's assertions, the purported specific variations of the App Store per device are not intrinsically linked to the online platform itself, which is fundamentally comparable across all five devices on which it is made available. Secondly, Apple offers identical distribution and support services to recipients of its service, irrespective of the device used to access the online platform'.

The DSA also employs the terms 'very large online platform' and 'very large search engine'. According to Article 33(1) and (4) of the DSA, these terms denote online platforms and search engines with over 45 million monthly active users in the EU, recognised by the European Commission as having such status through an administrative decision. To date, the European Commission has issued decisions regarding the designation of such entities in relation to the following entities [12].

The primary focus of the DSA is promoting a transparent and safe online environment with accountable actors that is also resilient to hybrid threats [9].

The Digital Services Act mandates that providers of very large online platforms, as well as very large online search engines ('VLOSES'), undertake comprehensive risk assessments and implement strategies to mitigate systemic risks that may arise from the design, operation, or usage of their services. Such risks encompass adverse impacts on civic discourse and electoral processes, as well as threats to public security, including significant interference by malicious state actors, particularly during electoral periods. The DSA includes a crisis response mechanism that can be activated in extraordinary circumstances posing a severe threat to public security or public health within the European Union or substantial parts thereof. To enhance this mechanism, the European Commission and the nationally designated competent authorities, known as Digital Service Coordinators, have developed a voluntary framework for addressing incidents arising under the DSA. Furthermore, Digital Service Coordinators have taken steps to safeguard the integrity of elections, which include organising electoral roundtables and conducting stress tests [20]. Together with the Regulation on Political Advertising (Regulation 2024/900) [46], the DSA represents a vital component of the broader initiative aimed at protecting democracy and the integrity of democratic processes, which are vulnerable to targeting by hostile actors through the use of digital tools and social media.

According to Decarolis et al. [16], the provisions set forth in the Digital Services Act focus on several pivotal issues regarding the obligations of digital platforms. Firstly, digital platforms are required to provide users with relevant and comprehensible information about the advertisers they encounter and the rationale behind their targeting. Under the regulations outlined in Article 26 of the DSA, all online platforms that display advertisements on their interfaces have a duty of transparency to ensure that service recipients can identify various critical aspects of each advertisement. This includes acknowledging content as an advertisement, the primary criteria used to determine the audience, and available options for modification. In addition to the transparency obligations, Article 26 of the DSA also effectively prohibits personalised advertising that utilises profiling based on special categories of data. In contrast, Article 28 of the DSA bans advertising based on profiling in general when the recipient is a minor.

Additional provisions of the Digital Services Act concentrate on the methodologies, systems, and techniques employed by online platforms to influence consumers' purchasing decisions. According to Article 3 of the DSA, a "recommendation system" is defined as a partially or wholly automated mechanism utilised by an online platform to suggest or prioritize particular information on its online interface. This definition encompasses the determination of the relative order or visibility of information presented as a consequence of a search conducted by the service recipient or through other means. Furthermore, Article 25 of the DSA specifically addresses "dark patterns," which denote subtle techniques embedded within the design or organisation of interfaces to guide and persuade consumers towards making specific decisions. Additionally, Article 27 of the DSA seeks to enhance the transparency of recommendation systems by identifying the type and visibility of commercial information presented to specific users of the online platform. It is also mandated that very large online platforms provide service recipients with the capacity to alter the parameters of the information presentation that are based on objective criteria, such as price, rather than on online profiling.

In Section 5 of the Digital Services Act, the legislator of the European Union has delineated additional obligations for very large online platforms. These obligations encompass, among others, the identification, analysis, and assessment of any significant systemic risks at least on an annual basis, as stipulated in Article 34(1)(a) of the DSA. Such risks include the proliferation of illegal content, as indicated in Article 34(1)(a) of the DSA, as well as any actual or anticipated adverse consequences for the exercise of various fundamental rights, civil discourse, electoral processes, and public security. This is particularly relevant concerning matters such as gender-based violence, the safeguarding of public health, and the protection of minors, along with severe negative ramifications for an individual's physical and mental well-being, as outlined in Articles 34(1)(b)–(d) of the DSA. It is imperative for VLOPS to meticulously consider how their content moderation systems, recommendation systems, and advertising selection and display systems influence these systemic risks, as per Article 34(2) of the DSA. The assessment encompasses the impact of intentional manipulation of their services, including inauthentic usage or automated exploitation, as well as the amplification and potential rapid and extensive dissemination of illegal content and information that is inconsistent with their terms and conditions. Furthermore, the assessment addresses specific regional or linguistic considerations, particularly when pertaining to a specific Member State. Additional obligations entail the implementation of risk mitigation measures, as prescribed in Article 35 DSA, and the conduct of annual independent audits, as dictated by Article 37 DSA. Recommendation systems and online advertising are subject to specific transparency obligations, as specified in Article 39 DSA. The legislator of the European Union has also mandated the establishment of a compliance function that operates independently from operational functions, consisting of one or more compliance officers, including the head of the compliance function. This compliance function must possess sufficient authority, status, and resources, along with access to the management body of the provider of the very large online platform or the very large online search engine, to oversee the compliance of the provider with the DSA, as stated in Article 41. Moreover, self-regulation plays a significant role in codes of conduct, marked in Article 45 of the DSA and Article 46 of the DSA concerning online advertising, as well as crisis protocols, as indicated in Article 47 of the DSA.

The European Union legislator has clearly delineated the fundamental obligations that VLOPS are mandated to fulfil concerning oversight by third parties in relation to compliance with the Digital Services Act. In this context, it is imperative to emphasise the necessity for VLOPS to ensure that notifications submitted by trusted flaggers are processed and adjudicated without undue delay, as stipulated in Article 22(1) of the DSA. Furthermore, VLOPS providers are obligated to facilitate collaboration with organisations conducting audits, thereby ensuring that these audits are conducted in an effective, efficient, and timely manner. This mandate encompasses, among other things, granting access to all relevant data and premises, as well as responding to both oral and written inquiries, as outlined in Article 37 of the DSA. According to Laux et al. [35], vetted researchers, characterised as academics possessing 'proven expertise' in relevant fields, shall be granted access to data upon request by a Digital Service coordinator to investigate the systemic risks specified in Article 34(1) of the DSA. Additionally, VLOPS are obligated to appoint compliance officers tasked with monitoring their adherence to the DSA, as mandated by Article 41 of the DSA. In accordance with the principle of transparency, providers of VLOPS that display advertisements on their online platforms are required to publicly disclose the information referenced in Article 39(2) of the DSA for the entire duration during which an advertisement is presented, as well as for one year following the final presentation of the advertisement on their online platforms, as stipulated in Article 39(1) of the DSA.

5.5 Conclusions

Digital mobility services are delivered within an ecosystem that encompasses digital platforms. Consequently, these services engage in legal relations pertaining to the subject matter of specific types of digital services. The market position of digital platforms may lead to competitive distortions in specific markets. This situation arises not only from the behaviors exhibited by digital platforms as stakeholders in the ecosystem, but also from the structure of the markets in which they operate. EU law has modified the legal framework to mitigate the market position of digital platforms, as stipulated by the Digital Markets Act. The implementation of EU competition law provisions, along with competition law precedents, does not negate the enforcement of the DMA, and vice versa. Under this regulation, the European Commission has identified a total of 24 gatekeepers that provide fundamental platform services. The DMA imposes several obligations on gatekeepers, including, but not limited to, providers of online intermediation services, many of which directly pertain to online intermediation services. Moreover, these obligations particularly establish enforceable rights for competitive online intermediation services. Thus, the DMA aims to enhance the differentiation of the conditions under which products and services are offered, thereby improving price competition for the benefit of end users and fostering competition among competing online intermediation services of the gatekeeper. Additionally, the regulation establishes several rights for commercial users of online intermediation services, including the right to market their products and services through competing online intermediation services and their own sales channels without restrictions. These users also possess the right to report irregularities to the Commission under the remedies provided for in the DMA. If gatekeepers violate their obligations under the DMA, the affected parties may pursue civil actions for injunctions (including interim measures) before national courts in the EU and, where applicable, initiate actions for damages. The Digital Services Act also pertains to the activities of digital platforms within the ecosystem concerning the provision of certain digital mobility services. An entity that provides the second type of digital mobility services may fall under the scope of the DSA. The third, fourth, and fifth types of digital mobility services, as outlined in Chap. 5, are offered through applications available in app stores. The operation of such app stores should also be regarded as an "online platform" within the definition provided in section 3(i) of the DSA. The DSA addresses several principal issues regarding the obligations of digital platforms. First, digital platforms are required to furnish users with pertinent and clear information about the advertisers they encounter and the reasons for being targeted. According to the stipulations of section 26 of the DSA, all online platforms displaying advertisements on their interfaces bear a transparency obligation to guarantee that service recipients can identify various key aspects of each advertisement. The DSA also prohibits personalised advertising that involves profiling based on special categories of data, and section 28 of the DSA specifically forbids profiling advertising where the recipient is a minor. Additional provisions of the DSA concentrate on the methods, systems, and techniques employed by online platforms to persuade consumers to make purchases.

References

1. Andriychuk, O.: Shifting the digital paradigm: towards a sui generis competition policy. Comput. Law Secur. Rev. **46**, 105733 (2022). https://doi.org/10.1016/j.clsr.2022.105733
2. Banaszczyk, Z.: System Prawa Prywatnego, t. 1. In: Safjan, M. (2007)
3. Cabanelas, P., Parkhurst, G., Thomopoulos, N., Lampon, J.F.: A dynamic capability evaluation of emerging business models for new mobility. Res. Transp. Bus. Manag. **47**, 100964 (2023). https://doi.org/10.1016/j.rtbm.2023.100964
4. Caillaud, B., Jullien, B.: Chicken and egg: competition among intermediation service providers. RAND J. Econ. **34**(2), 309–328 (2003)
5. Colomo, P.I.: Protecting the 'law' in competition law. J. Eur. Competition Law Pract. **11**(7) (2020). https://doi.org/10.1093/jeclap/lpaa074
6. Comeing, I., Klaser, K., Pinar, L.D.: The paradox of (inter)net neutrality: an experiment on ex-ante antitrust regulation. Technol. Forecast. Soc. Change **175**, 121405 (2022). https://doi.org/10.1016/j.techfore.2021.121405
7. Commission: Commission Decision of 25.4.2023 Designating Booking.com as a Very Large Online Platform in Accordance with Article 33(4) of Regulation (EU) 2022/2065 of the European Parliament and of the Council C(2023) 2748 Final (2023a). https://digital-strategy.ec.europa.eu/en/policies/list-designated-vlops-and-vloses. Accessed 1 Apr 2025
8. Commission: Commission Decision of 25.4.2023 Designating Booking.com as a Very Large Online Platform in Accordance with Article 33(4) of Regulation (EU) 2022/2065 of the European Parliament and of the Council C(2023) 2726 Final (2023b). https://digital-strategy.ec.europa.eu/en/policies/list-designated-vlops-and-vloses. Accessed 1 Apr 2025
9. Commission, 2025a: Communication from the Commission to the European Parliament, the Council, the European Economic and Social Committee and the Committee of the Regions on ProtectEU: A European Internal Security Strategy, 1.04.2025, COM(2025) 148 Final (2025)
10. Commission, 2025b: https://digital-markets-act.ec.europa.eu/gatekeepers_en (2025). Accessed 2 Apr 2025
11. Commission, 2025c: https://digital-markets-act-whistleblower.integrityline.app/?lang=en (2025). Accessed 1 Feb 2025
12. Commission, 2025d: https://digital-strategy.ec.europa.eu/en/policies/list-designated-vlops-and-vloses (2025). Accessed 2 Apr 2025

13. Consolidated Version of the Treaty on the Functioning of the European Union (TFEU) [2012] OJ C326/26
14. Council Regulation (EU) No 833/2014 of 31 July 2014 Concerning Restrictive Measures Given Russia's Actions Destabilising the Situation in Ukraine, OJ L 229, 31.7.2014, pp. 1–11
15. Crémer, J., Crawford, G.S., Dinielli, D., Fletcher, A., Heidhues, P., Heine, H., Schnitzer, M., Scott Morton, F.M., Seim, K.: Fairness and Contestability in the Digital Markets Act Digital Regulation Project, Policy Discussion Paper No. 3, 3.06. 2021. https://tobin.yale.edu/sites/default/files/2022-11/Digital%20Regulation%20Project%20-%20Fairness%20and%20Contestability%20-%20Discussion%20Paper%20No%203.pdf. Accessed 11 Apr 2025
16. Decarolis, F., Li, M.: Regulating online search in the EU: from the android case to the digital markets act and digital services act. Int. J. Ind. Organ. **90**, 102983 (2023). https://doi.org/10.1016/j.ijindorg.2023.102983. Accessed 11 Apr 2025
17. Directive (EU) 2015/1535 of the European Parliament and of the Council of 9 September 2015 Laying Down a Procedure for the Provision of Information in the Field of Technical Regulations and of Rules on Information Society Services (Codification) (Text with EEA Relevance) OJ L 241, 17.9.2015, pp. 1–15
18. Directive (EU) 2018/1972 of the European Parliament and of the Council of 11 December 2018 Establishing the European Electronic Communications Code (Recast) (Text with EEA Relevance), PE/52/2018/REV/1, OJ L 321, 17.12.2018, pp. 36–214
19. Dobkowski, J.: Struktura interesu publicznego a zasady rozdzielenia odpowiedzialności publicznoprawnej w administracji. In: Ura, E. (ed.) Jednostka, państwo, administracja—nowy wymiar. Rzeszów (2004)
20. DSA Elections Toolkit for Digital Services Coordinators 2025. https://digital-strategy.ec.europa.eu/en/library/dsa-elections-toolkit-digital-services-coordinators. Accessed 11 Apr 2025
21. Eisenmann, T.: Managing proprietary and shared platforms. Calif. Manage. Rev. **50**(4), 31–53 (2008)
22. Eisenmann, T.R., Parker, G., Van Alstyne, M.: Platform envelopment. Strateg. Manage. J. **32**(12), 1270–1285 (2011). https://doi.org/10.1002/smj.935
23. Ellison, G., Fudenberg, D.: Knife-edge or plateau: when do market models tip? Q. J. Econ. **118**, 1249–1278 (2003)
24. European Commission: Communication 'Shaping Europe's Digital Future', COM(2020) 67 Final, 8 (2020a)
25. European Commission: Digital Services Act package: Ex Ante Regulatory Instrument for Large Online Platforms with Significant Network Effects Acting as Gate-Keepers in the European Union's Internal Market. Inception Impact Assessment (2020b). https://ec.europa.eu/info/law/better-regulation/have-your-say/initiatives/12418-Digital-Services-Act-package-ex-ante-regulatory-instrument-of-very-large-online-platforms-acting-as-gatekeepers. Accessed 11 Apr 2025
26. European Commission: New Competition Tool. Inception Impact Assessment (2020c). https://ec.europa.eu/info/law/better-regulation/have-your-say/initiatives/12416-New-competition-tool. Accessed 1 Apr 2025
27. European Commission: Proposal for a Regulation on Contestable and Fair Markets in the Digital Sector (Digital Markets Act), COM(2020) 842 Final (2020d)
28. Ezrachi, A.: EU Competition Law Goals and the Digital Economy. Oxford Legal Studies Research Paper No. 17/2018. https://papers.ssrn.com/sol3/papers.cfm?abstract_id=3191766. Accessed 12 Feb 2025
29. Filipek, J.: Sprawa sytuacji prawnej i stosunku prawnego w prawie administracyjnym. In: Filipek, J. (ed.) Jednostka w demokratycznym państwie prawa. Bielsko-Biała (2003), s. 191
30. Genzlinger, F., Zejnilovic, L., Bustinza, O.F.: Servitisation in the automotive industry: how car manufacturers become mobility service providers. Strateg. Change **29**(2), 215–226 (2020)
31. Goettel, M.: 1.2. Elementy stosunku cywilnoprawnego. In: Goettel, M. (ed.) Prawo cywilne. Zarys wykładu. Warszawa (2022)
32. Hasselwander, M.: Digital platforms' growth strategies and the rise of super apps. Heliyon **10**(5), e25856 (2024). https://doi.org/10.1016/j.heliyon.2024.e25856

33. Judgment of the Court (Third Chamber) of 15 September 2016 Tobias Mc Fadden v Sony Music Entertainment Germany GmbH, Request for a Preliminary Ruling from the Landgericht München I, C-484/14, ECLI:EU:C:2016:689
34. Julià-Barceló, R., Koelman, K.J.: Intermediary liability: intermediary liability in the e-commerce directive: so far so good, but it's not enough. Comput. Law Secur. Rep. **16**(4), 231–239 (2000). https://doi.org/10.1016/S0267-3649(00)89129-3
35. Laux, J., Wachter, S., Mittelstadt, B.: Taming the few: platform regulation, independent audits, and the risks of capture created by the DMA and DSA. Comput. Law Secur. Rev. **43**, 105613 (2021). https://doi.org/10.1016/j.clsr.2021.105613
36. Motta, M., Peitz, M.: Intervention Triggers and Underlying Theories of Harm: Expert Advice for the Impact Assessment of a New Competition Tool. Publications Office of the European Union, Luxembourg (2020). https://competition-policy.ec.europa.eu/system/files/2021-03/kd0420575enn.pdf. Accessed 1 Apr 2025
37. Noe, T., Parker, G.: Winner take all: competition, strategy, and the structure of returns in the Internet economy. J. Econ. Manage. Strategy **14**, 141–164 (2005)
38. Noerr: https://www.noerr.com/en/insights/relevance-of-the-dma-for-online-intermediation-services (2025). Accessed 1 Feb 2025
39. Parker, G., Van Alstyne, M.: Information complements, substitutes, and strategic product design. Workshop on Information Systems and Economics, Charlotte, NC. http://ssrn.com/abstract=249585 (2000). Accessed 21 Feb 2025
40. Parker, G., Van Alstyne, M.: Two-sided network effects: a theory of information product design. Manage. Sci. **51**, 1494–1504 (2005)
41. Perez-Moure, H., Lampon, J.F., Velano-Rodriguez, M.E., Rodríguez-Comesana, L.: Revolutionizing the road: how sustainable, autonomous, and connected vehicles are changing digital mobility business models. Eur. Res. Manag. Bus. Econ. **29**, 100230 (2023). https://doi.org/10.1016/j.iedeen.2023.100230
42. Regulation (EU) 2016/679 of the European Parliament and of the Council of 27 April 2016 on the Protection of Natural Persons with Regard to the Processing of Personal Data and on the Free Movement of Such Data, and Repealing Directive 95/46/EC, OJ L 119, 4.5.2016, pp. 1–88
43. Regulation (EU) 2019/1150 of the European Parliament and of the Council of 20 June 2019 on Promoting Fairness and Transparency for Business Users of Online Intermediation Services (Text with EEA Relevance), PE/56/2019/REV/1, OJ L 186, 11.7.2019, pp. 57–79
44. Regulation (EU) 2022/1925 of the European Parliament and of the Council of 14 September 2022 on Contestable and Fair Markets in the Digital Sector and Amending Directives (EU) 2019/1937 and (EU) 2020/1828 (Digital Markets Act) (Text with EEA Relevance), PE/17/2022/REV/1, OJ L 265, 12.10.2022, pp. 1–66
45. Regulation (EU) 2022/2065 of the European Parliament and of the Council of 19 October 2022 on a Single Market for Digital Services and Amending Directive 2000/31/EC (Digital Services Act) (Text with EEA Relevance), PE/30/2022/REV/1, OJ L 277, 27.10.2022, pp. 1–102
46. Regulation (EU) 2024/900 of the European Parliament and of the Council of 13 March 2024 on the Transparency and Targeting of Political Advertising, OJ L, 2024/900, 20.3.2024
47. Resolution of the Supreme Court of 21.11.2000, III CKN 1048/00, LEX No. 51872
48. Rochet, J., Tirole, J.: Platform competition in two sided markets. J. Eur. Econ. Assoc. **1**(4), 990–1029 (2003)
49. Teece, D.J.: Big tech and strategic management: how management scholars can inform competition policy. Acad. Manage. Perspect. **37**(1) (2023). https://doi.org/10.5465/amp.2022.0013?journalCode=amp. Accessed 1 Apr 2025
50. The Act of 23 April 1964—The Civil Code (Consolidated Text: Journal of Laws of 2024, Item 1061, as Amended)
51. Truyens, M., Van Eecke, P.: Liability of domain name registries: don't shoot the messenger. Comput. Law Secur. Rev. **32**(2), 327–344 (2016). https://doi.org/10.1016/j.clsr.2015.12.018
52. Walden, I.: Mine host is searching for a 'neutrality' principle! Comput. Law Secur. Rev. **26**(2), 203–209 (2010). https://doi.org/10.1016/j.clsr.2010.01.003

53. de Vasconcelos, L., Farago, F., Facin, A., Flechas, X., Silva, L.: From open business model to ecosystem business model: A processes view. Technol. Forecast. Soc. Change **194**, Article 122668. (2023) https://doi.org/10.1016/j. techfore.2023.122668
54. Arthur, W. B.: Competing technologies, increasing returns, and lock-in by historical events. Econ. J. **99**, 116–131 (1989)

Chapter 6
The Driver of an Autonomous Vehicle as a Recipient of Digital Mobility Services

6.1 Legal Status of a Driver

The driver associates the digital mobility service with the autonomous vehicle. Consequently, the provision of such services necessitates prior deliberation regarding matters related to the driver's legal standing and responsibilities.

The driver of an autonomous vehicle is part of the group receiving digital mobility services discussed in Chap. 4, along with the passenger and the entity overseeing the fleet or platoon of autonomous vehicles. By viewing the driver as the primary focus for identifying digital mobility services, we can categorise these services into three groups:

(1) services that directly impact the driver's ability to perform driving tasks;
(2) services that have an indirect influence on the driver's driving tasks;
(3) services that neither directly nor indirectly affect vehicle operation but are related to transport organisation, assisting the driver in fulfilling tasks for operating the autonomous vehicle.

The initial group of services, which directly affects the driver's task performance, involves assistance with journey planning and execution. Consequently, these services need to be factored in when evaluating the driver's adherence to driving regulations.

Within the categories of digital mobility services discussed in Chap. 4, one can also identify services that indirectly influence the execution of tasks related to vehicle operation (i.e., the second category of services). This category encompasses digital mobility services of the fourth type, which facilitate access to content (including voice and data) that, although not essential for planning and executing the journey, enhances the overall comfort of the travel experience. Conversely, the use of these services by the driver during vehicle operation (i.e., while the vehicle is in motion) necessitates an assessment of their legal admissibility. This assessment is undertaken from the perspective of potential risks to road safety arising from decreased concentration.

A. Krasuski, *Legal Challenges of a Transformation from a Product-Centric to a Service-Centric Business Model and Procedures*, Studies in Systems, Decision and Control 610, https://doi.org/10.1007/978-3-031-98901-8_6

The concern regarding the impact of external stimuli on a driver's concentration, particularly concerning in-vehicle entertainment, has been extensively investigated in scientific studies. The use of entertainment while engaged in driving-related tasks should be assessed from both the perspective of the content provided and the way it is employed.

In the initial instance, it is noteworthy to highlight the research findings of Ünal et al. [82], who observed that music can serve as a demanding and distracting stimulus while driving. Concurrently, they underscored the necessity for drivers to employ cognitive compensatory strategies designed to reduce the distracting effects of music. Furthermore, Brodsky [8] identified that high-tempo music correlates with increased speeds and a higher frequency of red-light violations during simulated driving scenarios.

In light of how entertainment is integrated within the vehicle, it is crucial to highlight that digital mobility services of this nature can take the form of applications accessible through terminal devices, such as mobile phones, while operating within the vehicle. In this regard, based on the research conducted by Haigney et al. [31], it was noted that engaging with a mobile phone while driving could have implications for safety margins that are not readily observable. Although drivers seemed to make efforts to mitigate this by reducing their speed, the heightened cognitive load associated with mobile phone usage was evident. It can be anticipated that this factor will diminish drivers' ability to respond to emergencies or other unforeseen increases in the demands of the driving task. The negative impacts were particularly pronounced when participants used a portable phone.

In contrast, should the entertainment application be executed directly on permanently installed devices within the vehicle, research conducted by Horberry et al. [35] indicated that the level of additional in-vehicle task performance (including actions such as tuning the radio or engaging in conversation) while driving may adversely impact driving performance in certain driving scenarios. This study revealed that the in-vehicle task of interacting with the entertainment system (and, to a lesser degree, engaging in a simulated hands-free mobile phone conversation) could influence metrics of driving performance, including the maintenance of speed and readiness to respond to unexpected hazards (for instance, decelerating when a pedestrian unexpectedly crosses the road in front of the vehicle).

The third category of digital mobility services concerns the organisation of transportation. These services do not directly or indirectly influence the operation of a vehicle. However, they play a significant role by providing a rationale for transportation, thereby enabling the driver to engage in activities related to operating an autonomous vehicle. This category includes digital mobility services of the first, second, and fifth types, as outlined in the description of digital mobility services in Chap. 4.

The matter concerning the driver warrants consideration within the framework of driving automation levels that encompass the driver's (human) involvement in the operation of an autonomous vehicle (driving automation levels 1–3), as well as levels 4 and 5, corresponding to high driving automation and full driving automation, respectively. The utilisation of autonomous vehicles operating at the 4th and 5th levels of automation necessitates inquiries regarding the recipient of digital mobility services tailored to the driver. This relates to digital mobility services of types three and four, as delineated in Chap. 5. This aspect prompts the necessity to analyse the legal subjectivity of an artificial agent acting as a driver in an autonomous vehicle.

In the definition of 'autonomous vehicle' as delineated in Article 65k of the Polish Road Traffic Act [2], the legislator, by utilising the term 'control systems', has positioned the systems that manage an autonomous vehicle on an equal level with the actual driver. The legislator has assigned a normative significance to the term 'actual driver' within the Polish Road Traffic Act [2], as evidenced in Article 2(21), which defines it as a person who operates a vehicle or a combination of vehicles, as well as an individual who supervises a group of pedestrians, rides on horseback, or guides animals either singularly or in herding. This definition is expansive, encompassing, among others, a natural person who executes all the functions necessary for the operation of a particular vehicle. Furthermore, the actual driver is classified as a 'participant in the traffic' as articulated in Article 2(17) of the Polish Road Traffic Act [2].

The distinction of the concept of a 'driver' as articulated in the Polish Road Traffic [2] Act is of substantial importance for several reasons. A driver is recognised as a legal entity for whom the legislator has established specific legal prerequisites pertaining to the operation of a vehicle. The legislator has delineated formal criteria that an individual must fulfil to obtain official authorisation to operate a vehicle, to whom the designation of a driver has been assigned, as evidenced by a specific document.

In addition to the definition of 'actual driver', the Polish Road Traffic Act [2], as stated in Article 2(22), delineates the definition of a 'driver', which signifies that an individual is authorised to operate a motor vehicle or moped. Consequently, the concept of a 'driver' pertains to the administrative authorisation requisite for operating a vehicle.

Consequently, the subject pertains to an individual who, in the estimation of a certain public administration authority, possesses the requisite skills, knowledge, age, and health condition that enable them to operate a vehicle, as substantiated by an appropriate administrative identification in the form of a 'driving license' document. The definition of a 'driver' encompasses subsets that include the definition of an 'actual driver' and vice versa; however, these concepts are not interchangeable. The collective of individuals fulfilling the criteria of the definition of a 'driver' may also satisfy the criteria of the definition of 'actual drivers'. Conversely, only a subset of individuals meeting the criteria of the definition of the 'actual driver' will fulfil the criteria of the definition of the 'driver'. It must be presupposed that there exists, in reality, a cohort of individuals operating vehicles without the necessary authorisation as mandated by law.

As digital mobility services are examined from the perspective of autonomous vehicles in operation, further analysis relates to the concept of an 'actual driver', namely a natural person who undertakes the tasks associated with operating a vehicle.

The actual driver is an individual who is legally responsible for:

(1) obtaining the right to operate vehicles,
(2) fulfilling obligations towards administrative bodies, road traffic control authorities, passengers, and other road users related to operating a vehicle in road traffic,
(3) meeting obligations concerning the admittance of the vehicle to road traffic, the fulfilment of which is verified by the vehicle registration certificate. These obligations primarily fall upon the vehicle's owner or an independent possessor. However, they also extend to the actual driver of the vehicle, albeit indirectly. As a result, the driver is prohibited from operating a vehicle that does not satisfy these requirements.

In regard to the operation of vehicles in road traffic, incidents caused by the actions of the driver may occur, for which the driver may be held accountable under criminal, civil, and administrative law.

In accordance with Polish law, the actual driver bears criminal liability in ticket proceedings conducted by the Police, as stipulated in Article 95 § 1 of the Code of Procedure in Petty Offences [78] and the Regulation concerning the amounts of fines imposed in the form of penalty tickets for specific types of petty offences [63], which pertain to violations of orders and prohibitions delineated in the Polish Road Traffic Act [2] regarding the operation of the vehicle operated by him. In Section II, Chapter 3 of the Polish Road Traffic Act, entitled 'Movement of Vehicles', the legislator articulates the obligations of a vehicle driver in relation to the operation of this vehicle. Additionally, the actual driver may also be deemed a perpetrator of offences against road safety, as regulated in Chapter XXI of the Polish Penal Code [79].

The actual driver may also bear liability for tortious acts in accordance with the provisions of the Polish Civil Code [77] regarding the operation of a motor vehicle, should he or she be classified as either an independent or dependent possessor. This is in conjunction with the fulfilment of additional liability requirements as articulated in Article 436 in conjunction with Article 435 of the Polish Civil Code [77]. Furthermore, the actual driver is subject to administrative sanctions, which may encompass the revocation of the driving licence, pursuant to the regulations detailed in Chapter 3, Part II of the Polish Road Traffic Act [2].

The definitions of "actual driver" and "driver" specified above relate to an autonomous vehicle, which, as defined in Article 65k of the Polish Road Traffic Act [2], is classified as a motor vehicle. Furthermore, in accordance with Article 2(33) of the Polish Road Traffic Act, a motor vehicle is characterised as a vehicle designed to operate at a speed exceeding 25 km/h, except for agricultural tractors.

Considering the definition of an 'autonomous vehicle' articulated in Article 65k of the Polish Road Traffic Act [2], it is essential to reference the automated driving system ('ADS') regarding the legal status of an autonomous vehicle. The ADS is responsible for managing the vehicle's functionalities, facilitating its steering, and allowing the vehicle to traverse a predetermined distance. In this context, the operation of the ADS ensures effective driving, even when a human engages in only a portion of the driving tasks. Consequently, an individual operating an autonomous vehicle assumes a dual role. Firstly, they serve as a role model for the ADS; the observation of the human driver's performance in executing driving functions aids in establishing a dataset for artificial intelligence training while also defining the behavioural rules incorporated within the artificial intelligence algorithms. Secondly, humans act as collaborators with the artificial intelligence, as they fulfil the steering functions of an autonomous vehicle at the third level of driving automation alongside the ADS. In principle, an ADS should possess the capability to operate an autonomous vehicle at the two highest levels of automation, specifically levels 4 and 5. Thus, it is imperative to ascertain whether an ADS, in its capacity as an artificial agent, can be recognised as a distinct legal entity on par with a human driver.

In light of the necessity for both the human driver and the automated driving system to display comparable driving competencies concerning an autonomous vehicle, and thus be capable of operating independently, it is imperative to ascertain whether both the automated driving system and the human driver can effectively utilise digital mobility services. This is particularly relevant within the framework of the third and fourth categories of these services as delineated in Chap. 4. The services classified under the third category are specifically tailored for the driver of an autonomous vehicle, while those under the fourth category are designed for the benefit of both the driver and the passenger. The assertion that an ADS, being an artificial construct, cannot qualify as an independent legal entity implies that the digital mobility services of the third and fourth categories may be exclusively directed towards human beneficiaries. As a result, such digital mobility services, particularly those belonging to the third category, would exclusively pertain to level 3 of automation. Furthermore, they would be excluded from applicability under automation levels 1–2, as these levels do not, by definition, take into account the capabilities of an artificial agent. Additionally, they would also be excluded from consideration under automation levels 4 and 5, as at these levels, the human driver does not assume the responsibility of carrying out driving functions.

The definitions of 'actual driver' and 'driver' relate to the definition of 'vehicle'. This concept is outlined in Article 2(31) of the Polish Road Traffic Act [2], which specifies that a vehicle is a means of transport specifically designed for motion on the road, as well as a machine or device intended for this purpose, excluding any device that merely supports movement. As a means of transport, a vehicle operates on the road in accordance with Article 4 item 2 of the Polish Act on Public Roads [3].

In conclusion, the matter of the legal status of ADS, which allows for the undertaking of tasks on equal terms with humans (as an actual driver), is significant for identifying the group of legally distinct recipients of digital mobility services.

At the international level of these considerations, it is essential to emphasise the regulations of the Convention on Road Traffic [14], which serve as a legal reference for the aforementioned considerations. Article 8, Section 1 of the Convention on Road Traffic [14] stipulates that every vehicle in motion must be operated by a driver. However, a literal interpretation of this provision does not necessitate that the driver be a human. Moreover, this assertion has been corroborated in the relevant literature on the subject [88]. The literal interpretation aligns with the functional interpretation, which is grounded in the overarching purpose of the Convention on Road Traffic [14]: to ensure road safety. The concept of autonomous vehicles is consistent with the objectives underlying the introduction of this Convention. In accordance with Article 8 of the Convention on Road Traffic [14], it is also unnecessary for the driver to remain in the vehicle while it is in motion. Therefore, it is permissible to recognise either a human or an artificial agent as the 'driver'.

Furthermore, as stipulated in Article 8, Section 5 of the Convention on Road Traffic [14], it is imperative for the driver to consistently maintain control over their vehicle. The precise wording of this provision denotes an obligation to exercise control over the vehicle, understood as utilising its control functions to ensure road safety. This implies that when an autonomous vehicle encounters other road users, the driver (represented as an artificial agent) must take any necessary precautions to guarantee their safety. The Convention on Road Traffic [14] does not require the driver to be physically present within the vehicle. Therefore, both a self-learning machine and a human operator situated outside the vehicle may be classified as 'drivers'. There is a need to consider initiatives aimed at introducing remotely controlled autonomous vehicles, particularly for commercial purposes, on public roads. In such scenarios, individuals controlling these autonomous vehicles (i.e., human operators) would typically be recognised as 'drivers'. A notable exception is the concept of the 'remote ADS assistant', as discussed in relevant literature, which envisions a human role outside the autonomous vehicle to support the ADS. This support encompasses tasks performed by the ADS, excluding the driving of the autonomous vehicle, such as route planning and monitoring traffic disruptions [12].

In factual and functional terms, it should be noted that the occupant of an autonomous vehicle operating at automation levels 4 and 5 ought not to be regarded as a driver, but rather as a passenger throughout the entirety of the period during which the vehicle is in motion. In instances of an autonomous vehicle functioning at automation level 3, the role of the human is twofold. Should the human assume control of the driving function in response to a request from the ADS, the human functions as the driver until such time as the vehicle is no longer in motion or the human relinquishes the driving function back to the ADS. During the duration in which the ADS is activated and prior to the human taking over the driving responsibilities from the ADS, the human's role is diminished to monitoring the ADS's behaviour and promptly responding to the ADS's request for immediate transition of driving functions. In the scenarios outlined above, the ADS assumes the role of the driver when it effectively manages the technology necessary to operate the autonomous vehicle at automation levels 3–5. Irrespective of who acts as the actual driver, diligence is imperative when supervising the autonomous vehicle [36].

The aforementioned argument regarding the absence of contraindications to conferring driver status upon an artificial agent, in accordance with Article 8, paragraph 1 of the Convention on Road Traffic [14], is not negated by Article 8, paragraph 4 of the said Convention. This provision stipulates that every operator of a motor vehicle must possess the requisite knowledge and skills necessary for its operation. Furthermore, the stipulation presented in Article 8, paragraph 4 of the Convention on Road Traffic [14] does not hinder the utilisation of machine learning by the artificial agent. In such instances, the artificial agent could fulfil the criteria outlined in this provision.

Article 8, paragraph 3 of the Convention on Road Traffic [14] delineates the necessary physical and mental fitness standards for drivers. It is noteworthy that artificial agents hold advantages over human operators, primarily due to their exemption from limitations imposed by factors such as age, health complications, or dependencies. These limitations may include movement disorders, visual and auditory impairments, as well as psychological conditions. Such issues may negatively impact the physical and mental capacities essential for ensuring safety while driving.

In Polish law, the ADS, which encompasses both material objects and intellectual property rights, does not possess a separate legal entity. The recognition of an artificial agent as a distinct legal entity for the purpose of designating a responsible party is presently unattainable, as an artificial agent is devoid of:

(1) awareness regarding the decisions rendered and their subsequent effects,
(2) comprehension of the implications of sanctions imposed as a consequence of legal infractions associated with the decisions rendered,
(3) the capability to grasp the rationale for compensation related to injury or damage inflicted by the conduct of an artificial agent,
(4) the capacity to comprehend the justification for penalties levied in connection with the commission of prohibited acts,
(5) the ability to understand and assert the rights to which the perpetrator (human) is entitled within the legal system.

The current legal status of an artificial agent, defined as a robot employing artificial intelligence capabilities, exists as either an implicit moral agent or an explicit moral agent, and remains provisional. Changes to this status are contingent upon several factors: (1) the advancement of engineering technology in the realm of self-learning software development, which facilitates the creation of tools that enhance the integration of the artificial agent with its environment and the human beings inhabiting it; (2) the evolution of the environment in which the artificial agent operates, thereby increasing its capacity to exhibit characteristics such as consciousness, free will, and the intention to pursue its own objectives.

From a legal perspective, to appropriately regulate the distinct legal status of an artificial agent, the legislator must consider the following factors. As the sophistication of the automated driving system increases with higher levels of automation, there is a transition from the status of an actual driver (human) to that of a passenger. When analysing the functionalities of level 4, representing high automation, and level 5, corresponding to full automation, it becomes evident that there is no 'ultimate decision-making authority' in the form of a human acting as a driver. The delineated levels of automation, designed to ensure vehicle autonomy, result in the decisions made by the system at levels 4 and 5 becoming overriding decisions.

In the process of formulating prospective legal regulations pertaining to autonomous vehicles, lawmakers must consider the potential inability to exercise control over the vehicle's operational functions from the viewpoint of a possible infringement on the freedom of movement guaranteed in the Constitution of the

Republic of Poland [80]. This freedom includes the right to traverse not only within the territorial scope of Poland but also beyond its borders. It is imperative to understand freedom of movement, with the caveats delineated below, in such a manner that it encompasses the right to intervene in the driving process while the ADS is in operation, specifically within the automation levels of 3–5. At a bare minimum, such intervention should facilitate an alteration in driving style through the vehicle's cessation in a safe and legally permissible location. Neglecting this possibility would result in an infringement of liberty for the occupant of the vehicle, thereby contravening Article 41 of the Constitution of the Republic of Poland [80]. Furthermore, adherence to a passenger's request to halt the vehicle in an authorised area aligns with the established practices for utilising transport services in both public and private transportation contexts.

In the context of implementing the highest two levels of automation, the owner (driver) of the vehicle, despite being present within the vehicle, loses situational awareness, thereby diminishing the fundamental basis for exercising 'dominance' or 'power' [26]. Paradoxically, the interpretation of Article 52, Sections 1 and 2 of the Constitution of the Republic of Poland [80], considering the right of a human to intervene in the process of controlling the vehicle at these two highest levels of automation, may consequently result in a reduction in the safety of autonomous vehicle movement when a human invokes this right. In the long term, the implementation of automation levels 4 and 5, which entirely exclude humans from the driving role, may also contribute to a decline in driving skills [70].

When formulating future regulations, it is imperative for the legislator to analyse the fundamental question of whether an individual possesses the physical and mental capacity to assume the role of steering an autonomous vehicle, a function previously executed by the ADS, without compromising the safety associated with the movement of such a vehicle. Among the advantages of delegating the steering of an autonomous vehicle to an artificial agent are features that diverge significantly from human capabilities, including, but not limited to, accelerated decision-making, the absence of fatigue, enhanced visual acuity at all times of the day and in various weather conditions, and the absence of diseases and addictions that adversely impact driving performance. Consequently, the right of an individual to assume control of the vehicle's steering functions from an artificial agent at the two highest levels of automation should be subjectively limited. In this context, it is also necessary to ascertain whether this right should be implemented by:

(1) the previous driver (human),
(2) a passenger who is not the driver (human),
(3) another person outside the vehicle (in the scope of remote control of an autonomous vehicle).

6.2 The Influence of Digital Mobility Services on the Decision-Making Challenges Faced by Operators of Autonomous Vehicles

The digital mobility services discussed in Chap. 4 include those that facilitate the operation of (semi and fully) autonomous vehicles. Therefore, it is essential to analyse how these services impact the decision-making dilemmas associated with autonomous vehicles.

A decision-making dilemma arises when a driver, confronted with an event during the operation of a vehicle, must select a course of action from multiple options, each leading to potential violations of equivalent legal, social, or moral interests. Given the unique characteristics of autonomous vehicles, it is imperative for legislators to establish future legal requirements that align with societal expectations regarding how artificial agents navigate such dilemmas. Conversely, it is plausible that certain traffic-related dangers stem from intrinsic characteristics and cannot be mitigated (for example, the diverse interactions among road users with varying levels of protection). In its resolution on January 8, 2019 (II KK 145/18) [65], the Polish Supreme Court articulated that an individual whose life is endangered by the serious misconduct of another road user cannot be expected to passively await rectification of the hazardous manoeuvre, as such rectification is not always feasible due to objective factors.

In its earlier judgments dated 16 July 1975 (I KRN 79/75) [74] and 29 September 1993 (III KRN 127/93) [40], the Supreme Court articulated similar perspectives, culminating in the assertion that the principle of limited trust cannot be construed so expansively as to render drivers liable for every circumstance arising on the road (i.e., effect). It is pertinent to recall that in the resolution of the full bench of the Criminal Chamber dated 28 February 1975 (V KZP 2/74) [64], the Supreme Court conveyed the opinion that 'a person driving a vehicle has the right to expect adherence to traffic safety rules by other road users, provided that their personal characteristics, specific behaviours, or any other exceptional situation, justified by life experience (e.g., passing a bus stationary at a bus stop, thereby obstructing visibility or driving in a pedestrian crossing zone), do not give rise to the expectation that such rules will be disregarded.'

The principle of limited trust in road traffic is interpreted restrictively within the literature on criminal law [9, 45, 61, 72]. Threats may not solely arise from the vehicle control system, instead, they may be inherent to road traffic itself. Consequently, these threats can be attributed to the complexity and diversity of potential scenarios within the 'road traffic system'. This interpretation appears especially plausible when considering the influence of driving speed, which is arguably a necessary causal factor (in the scientific sense) in the occurrence of road accidents in the majority of instances. Furthermore, 'normal' driving speed is a fundamental aspect of contemporary understandings of the 'road traffic system' and significantly influences its characteristics [68]. In its ruling dated 20 October 2022 (IV KK 341/21) [39], the Supreme Court stated that the principle of maximising safety cannot be interpreted as a behavioural rule that guarantees, with 100% certainty, the absence of negative outcomes. Such a rule would often necessitate reducing speed to zero and bringing the vehicle to a complete stop.

The assessment of decision-making dilemmas should be anchored in constitutional principles related to the rights and freedoms of citizens, particularly concerning the protection of lives (see Article 38 of the Polish Constitution [80]) and the safeguarding of citizens' safety (see Article 5 of the Polish Constitution [80]). In the context of the fundamental right to life and the assurance of safety, the transition from human control to machine control of vehicles does not modify any matters of fundamental significance. Specifically, no "targeted" interference with life or safety would be considered practically unjustifiable given the paramount importance of these rights (see Supreme Court judgment of 19 February 1974, Rw 53/74 [75]). Although each control decision is dictated by programming algorithms under specific boundary conditions, and thus is ultimately predetermined rather than random, it does not represent a specification of a particular course of action.

The programming of the autonomous driving function dictates the aspects that must be considered, allowing the driver to choose from various alternatives the option that completely avoids damage or results in the least harm. The significant added value of autonomous vehicle control lies in its ability to consider alternative actions based on situational parameters. However, it is essential to note that no control decisions are made within the programming; rather, only abstract criteria are established for specific control decisions. Consequently, it becomes clear that, ultimately, the operation of machines in road traffic continues to involve abstract risks. In the context of fundamental rights, it should therefore be regarded no differently than the operation of a vehicle by a human, which also carries inherent risks.

Digital mobility services that directly support human drivers operating semi-autonomous vehicles may include features such as locating the nearest electric vehicle charging points along the route or suggesting efficient driving techniques that ensure minimal energy or fuel consumption when the vehicle is powered in this way. Undoubtedly, these services aim to provide drivers with information about the actions taken while controlling the vehicle. Consequently, this information can significantly influence the decisions made by drivers, encouraging them to perform specific manoeuvres on the road based on the recommendations from these services. Assuming digital mobility services that assist in driving a vehicle are not mandatory equipment, the choice to acquire them, and therefore the necessity to use them, lies with the driver. It is up to the driver to determine the extent to which he or she chooses to depend on the information obtained via the service when making decisions relevant to the operation of the vehicle.

Assuming that the use of digital mobility services is not legally restricted, this allows a driver to access an unlimited number of digital mobility services designed to assist them in performing driving tasks. Consequently, there is a risk of accumulating an overwhelming amount of information that the driver must manage while operating the vehicle. This situation may lead to distractions. Furthermore, distractions and feelings of uncertainty while driving can also be influenced by discrepancies in the information provided to the driver on identical topics from various digital mobility services.

A human driver must therefore independently and at their own risk assess which information is useful and upon which they will base their decisions regarding the vehicle. Driver distraction may limit their ability to exercise caution on the road, and in certain instances as provided by law, also necessitate the exercise of special caution (Article 3 of the Polish Road Traffic Act). The legislator has defined 'special caution' in Article 2(22) of the Polish Traffic Act as a form of caution entailing increased attention and the adaptation of a traffic participant's behaviour to conditions and situations that evolve on the road, allowing for an appropriately swift reaction. The concept of 'special caution' has been extensively analysed in the case law of Polish common courts. Pertaining to the matter at hand, it is imperative to reference the judgment of the Supreme Court dated 19 February 1974 (Rw 53/74). In this judgment, the court delineated the concept of 'special care' as 'prudent driving of a motor vehicle, consisting in undertaking (on the part of the driver) such actions that, in accordance with the art and technique of driving motor vehicles, are objectively necessary to ensure maximum safety in road traffic, as well as refraining from actions that could reduce such safety'. Consequently, the court indirectly addresses the proliferation of digital mobility services and the information they generate, which could diminish the driver's concentration, thereby negatively impacting road safety.

When the vehicle operates in fully autonomous mode, specifically high automation classified as level 4 and full automation categorised as level 5, the necessary decisions regarding the demand for digital mobility services and the utilisation of the information gathered from these services will be determined by the ADS. More specifically, the decision-making process will be influenced by the programming of the control algorithms. These algorithms facilitate a comprehensive analysis of the data acquired through the machine learning process. Such data is sourced from various origins, including the environment in which the autonomous vehicle traverses.

According to Echeto et al. [18], vehicles, analogous to ants, transmit signals similar to pheromones directed towards other automobiles, as exemplified by perceptual signals such as braking lights and variations in speed. Vehicles possess the capacity to interpret signals emitted by other vehicles and modify their speed and direction accordingly. This mode of communication is primarily governed by mandatory regulations. The Polish Road Traffic Act [2] delineates various forms of traffic communication. This legislation stipulates the utilisation of specific visual and auditory signals by a moving vehicle. This includes audible signals, such as the horn, turn signals, or hazard lights, utilised under their legal authority. For instance, the activation of hazard warning lights indicates a warning of potential danger. Under specific circumstances, a driver is required to engage the hazard warning lights, such as when a motor vehicle (or trailer) is halted due to a malfunction or accident on a motorway or expressway. If the vehicle lacks hazard warning lights, the obligation to activate the position lights remains unaffected by the requirement to deploy a reflective warning triangle (Article 50 sec. 2(1)(d) and Article 50 sec. 2(2) of the Polish Road Traffic Act [2]).

Furthermore, the stipulations regarding the towing of a motor vehicle specify that the towed vehicle, except for a motorcycle, must be marked on the left rear with a reflective warning triangle. Additionally, during periods of reduced visibility, the side lights must also be activated. Alternatively, instead of using a reflective triangle, the towed vehicle may emit yellow flashing signals in a manner that is visible to other road users (Article 30 sec. 1(5) of the Road Traffic Act [2]).

The use of sound signals or indicators is mandated or permitted by the Polish Road Traffic Act [2]. In areas outside of built-up zones, during instances of fog, the operator of a vehicle is required to emit short sound signals when overtaking or passing another vehicle (Article 30 sec. 1(1)(b) of the Road Traffic Act). Furthermore, the operator of an emergency vehicle may, provided that utmost care is exercised, deviate from the regulations governing vehicle traffic, stopping, parking, and compliance with road signs and signals, solely when the vehicle simultaneously emits both light and sound signals (Article 53 sec. 2(2) of the Road Traffic Act). The implementation of indicators is additionally required and advised by the Polish Road Traffic Act. The operator of a vehicle is obligated to signal in advance and clearly with the indicator their intention to change direction or lane, and must cease signalling immediately after the manoeuvre has been completed (Article 23(5) of the Road Traffic Act). Should the vehicle lack indicators, the legislator imposes an obligation to extend a hand in the direction of the intended lane change or directional shift (Article 23(5a) of the Road Traffic Act). The legislator associates the utilisation of an indicator with the establishment of additional responsibilities. A vehicle operator approaching a designated bus (trolleybus) stop within a built-up area is required to reduce speed and, if necessary, come to a stop to facilitate the bus (trolleybus) operator's re-entry into traffic, particularly if the latter signals their intention to change lanes or enter the carriageway from a designated bay (Article 561 of the Polish Road Traffic Act).

The data used for machine learning can also be generated from digital mobility services. The extent to which information derived from digital mobility services is integrated depends on its application in the decision-making process of Autonomous Driving Systems (ADS) concerning the movement of autonomous vehicles. This integration will be influenced by the programming of the algorithms. Consequently, it is imperative to emphasise that decisions made by ADS in road scenarios, particularly those involving complex decision-making dilemmas, such as the necessity to 'sacrifice' one good for another, should be regarded as well-considered.

The programming of algorithms may consequently influence the prioritisation of certain road users [54] over others, thereby shaping the risks associated with the behaviour of an autonomous vehicle on the road. This prioritisation is related to the moral dimensions of the decisions rendered. Morality serves as a reference point for the concept articulated in the literature regarding the development of a real-time system aimed at estimating potential damage in road accidents. Its objective is to assist the computer controlling an autonomous vehicle (ADS) in determining the least harmful consequences of a road accident when the ADS deems the accident to be unavoidable, by analysing all potential damages associated with each possible road incident [86].

It is essential to initiate a discussion concerning the ethically acceptable distribution of risk related to the decisions made by Autonomous Driving Systems (ADS) and the operation of autonomous vehicles [46]. Consequently, it is crucial for legislators to establish regulations that clearly outline the criteria for decision-making through algorithms [20]. Any violation of the established algorithms by a driver should be considered a separate prohibited act. In engaging in such a discussion, ethical considerations should also be accounted for. It is important to highlight the distinction between ethics and morality, as Levinas [49] elucidates, ethics pertains to the relationship with another individual in their individuality, whereas morality is conceptual in nature. Ethics comprises a set of beliefs that society imparts to its members to encourage positive-sum interactions while discouraging negative-sum interactions [16]. A more comprehensive definition of ethics may be described as a collection of moral principles and norms (or forms of conduct) provided by external sources (such as a community or social environment) to which an individual belongs, thereby distinguishing between 'good and bad' or 'right and wrong'. Applied ethics encompasses numerous specific domains, including machine ethics, technology ethics (technoethics), cyber ethics, and digital ethics [56]. In the realm of artificial intelligence, ethics delineate the ethical and moral obligations and responsibilities of both the AI system and its creators [43].

6.3 Legal Challenges Associated with the Granting of Separate Legal Personality to an Artificial Agent in Relation to the Operation of an Autonomous Vehicle

The digital mobility services outlined in Chap. 4 relate to the direct or indirect use of autonomous vehicles. Consequently, it is essential to analyse the positions expressed in the literature regarding the potential granting of the status of a distinct legal entity to an artificial agent (i.e., ADS) in relation to the operation of an autonomous vehicle. The prevailing uncertainty surrounding the allocation of liability for incidents connected to the operation of an autonomous vehicle negatively affects the development and implementation of such vehicles [51, 60].

The Oxford Commission on Artificial Intelligence conducted an analysis of global attitudes toward the role of artificial intelligence in decision-making, utilising a comprehensive sample of 154,195 respondents across 142 countries [59]. The findings of the study indicate that automated artificial intelligence decision-making is perceived as potentially harmful, with North America exhibiting the highest concern at 47%, in contrast to significantly lower levels of apprehension reported in East Asia at 11%. This study employed data sourced from the 2019 World Risk Poll ('WRP'), ensuring that a minimum of 1000 respondents were surveyed within each country. A probability-based sampling methodology was utilised to represent the entirety of the population aged fifteen and over. Furthermore, attitudes toward artificial intelligence were examined across various demographic factors, including gender, education level, income, working hours, and attitudinal risk, among other variables elaborated upon in the study. The term 'trust' is defined as a psychological state encompassing the willingness of a trustee to accept vulnerability based on positive expectations regarding the trustor's intentions or behaviour [67]. Trust can influence or arise from trust-related behaviours (e.g., cooperation, risk-taking), serving as a motivating factor for action, although it is not a behaviour in itself. Ethics and trust are intricately connected, a high-trust environment fosters the development of ethical standards, while ethical behaviour contributes to the cultivation of trust [56].

The examination of the potential attribution of responsibility for incidents resulting from the operation of an autonomous vehicle directly to an artificial agent should be grounded in the contrasting perspectives articulated within the literature.

According to the initial perspective, when assessing criminal liability for a road traffic accident that occurs within the operational driving domain ('ODD'), responsibility should be assigned to the artificial agent (ADS) rather than to the occupant of the vehicle (i.e., the individual who acted as a passenger from the moment the vehicle was initiated and the individual who became a passenger following the activation of the ADS, contingent upon the level of automation that delegates the driving function directly to the ADS) [36]. Consequently, the artificial agent should be endowed with a separate legal entity. The existing literature on this matter indicates that the legal entity of an artificial agent does not need to replicate the legal entity of a human or the legal entity of legal persons. It should be precise, contextual, and restricted solely to those domains of artificial intelligence activity wherein the grant of subjectivity to the artificial agent is warranted by its social role [47]. Individuals ought to be held legally and administratively accountable for neglecting to install, maintain, and inspect artificial intelligence systems that facilitate autonomous driving, including the ADS.

According to an alternative perspective, an artificial agent cannot be held liable for incidents resulting from the operation of an autonomous vehicle, as the onus of responsibility should reside with individuals, specifically:

(1) manufacturers of Autonomous Driving Systems [17, 18, 22, 23, 25, 27, 30, 84, 85], or
(2) entities that maintain or knowingly contribute to the functionality of autonomous vehicles [52], or
(3) both independent and dependent owners of an autonomous vehicle [66].

In nations characterised by a significant degree of automation in transportation means, investigations have been undertaken concerning the societal acceptance of public transport systems utilising autonomous vehicles. For instance, a study conducted in China demonstrates that the perception of autonomous transportation is predominantly favourable, and autonomous vehicles exert a positive influence on users [41], aligning with the findings presented at the outset of this chapter.

In the context of Chinese law, the determination of human liability is based on the presence of an individual in the driver's seat of the vehicle, irrespective of their active engagement in the driving function. For instance, the authorities in the city of Shenzhen, China, promulgated a regulation governing liability for incidents resulting from the operation of an autonomous vehicle [69]. The regulation stipulates that if an individual occupies the driver's seat while the vehicle operates autonomously, the driver is held accountable under criminal, civil, and administrative law. Conversely, in the absence of an individual in the driver's seat, the independent possessor or user of the vehicle bears tort liability. Additionally, should latent defects in a vehicle lead to property damage or personal injury, the owner or user of the vehicle, upon settling the claim, is entitled to pursue a recourse claim against the manufacturer or seller of the vehicle, specifically concerning physical defects. It is important to note that the burden of proof rests with the driver (user) of the autonomous vehicle.

The perspectives and instances of regulations previously mentioned relate to a significant legal matter concerning the potential classification of an artificial agent as a distinct legal entity. The conceptualisation of artificial agents, particularly robots, raises inquiries about the anthropological distinctions between humans and machines [55]. Concurrently, in Japan, deliberations regarding societal development trends consider the coexistence of humans and artificial agents [44].

The matter concerning the legal subjectivity of artificial agents requires consideration within a broader ethical context, thereby raising the question of whether artificial agents are capable of exhibiting moral behaviour [21]. It seems that other road users and society at large are likely to interpret the actions of autonomous vehicles, along with the priorities established by their programmers, through an ethical lens.

The perspectives articulated in the literature concerning the legal qualification of artificial agents exhibit a notable division. Certain scholars evaluate the artificial agent based on the physical (visual) attributes characteristic of human beings. By positing that an artificial agent lacks these attributes, they classify it as a material object, essentially categorising it as a mere thing. A straightforward definition within this context was provided by Zech [87], who stated that robots are technical products. Conversely, Van den Hoven van Genderen [83], in his position paper, emphasises that an artificial agent, which embodies the characteristics of a robot, may be regarded as an object of a legal relationship. This perspective does not preclude the possibility of granting it the status of a subject within a legal relationship, as the legal nature is ultimately determined by human beings through the formation of legal relationships.

The consequence of regarding an artificial agent as the subject of a legal relationship is the assumption that the artificial agent lacks the status of a moral agent. An artificial agent cannot be attributed the status of a moral agent and, therefore, cannot possess legal rights and obligations.

The primary characteristic of a moral agent is its capacity to perform actions autonomously; that is, the actions of the artificial agent are not predetermined by a third party. This implies that the artificial agent undertakes these actions independently of human influence, meaning that a human being has no sway over the execution of these actions, the manner in which they are carried out, or the cessation of their performance [42].

An artificial agent could be regarded as an autonomous subject in the strictest sense of the term, provided there exists a rational justification for the actions of the artificial agent [73]. According to Bertolini [7], the critical issue at hand is the (self-) consciousness of the artificial agent. Therefore, insofar as the artificial agent fails to attain consciousness, it cannot be categorised as a moral agent or an autonomous entity in the strictest sense of the term.

An artificial agent, which is the subject of legal trading, lacks the characteristics of autonomy, including awareness, intention, and free will.

In light of the aforementioned considerations, a pertinent inquiry arises regarding the feasibility of designing a priori an autonomous system responsible for steering an autonomous vehicle through the application of suitable control algorithms. These algorithms must ensure that the autonomous steering system functions in a manner that encompasses not only universally applicable regulations, including those pertaining to conduct in road traffic, but also the ethical principles associated with such regulations. Ethical considerations concerning steering, braking, and accelerating an autonomous vehicle have been addressed by Lin et al. [50] and Goodall [28].

An autonomous vehicle is required to adhere to traffic laws, except in instances where compliance with such laws may conflict with the initial three laws established by Asimov [6], specifically:

(1) An autonomous vehicle must not collide with a pedestrian or cyclist;
(2) An autonomous vehicle must not collide with another vehicle unless evading such a collision would contravene the first law;
(3) An autonomous vehicle must not collide with any other object within the environment unless avoidance of such a collision would conflict with either the first or second law [27].

Consequently, it may be deemed acceptable to contravene traffic regulations in order to safeguard human life. This approach serves as a means of addressing the decision-making dilemma. Traffic regulations effectively embody deontological constraints, considering the straightforward nature of the stipulations. Motor vehicles are required to halt at red traffic signals and pedestrian crossings, adhere strictly to the designated speed limits, and refrain from crossing double solid lines, among other regulations. Nonetheless, individuals frequently perceive these laws as mere guidelines rather than unequivocal legal obligations.

The Council of Europe report "Legal Aspects of Automated Vehicles" [5] delineates the following ethical principles pertaining to the use of autonomous vehicles: transparency, fairness, and honesty; responsibility; safety and security; and privacy. Transparency necessitates that AI systems are accessible and comprehensible, allowing individuals to understand how these systems operate and produce results. Justice and fairness encompass non-discrimination, impartiality, consistency, and an acknowledgement of diversity and pluralism. Those who utilise AI systems must have the opportunity for remedies and redress.

For instance, the frequency with which human drivers stop at four-way intersections initially posed challenges for Google's autonomous vehicles, which awaited the cessation of motion of other vehicles [29]. Adherence to traffic regulations is generally assessed in relation to other objectives, such as safety, optimal traffic flow, or efficiency. These instances occur so routinely that it becomes difficult to argue that humans strictly observe traffic regulations as if they imposed absolute constraints on behaviour. For example, Google's autonomous vehicles have been programmed to exceptionally exceed the speed limit [37] when necessary to maintain traffic flow and adhere to scheduled travel times. If autonomous vehicles are designed to coexist with human drivers on the road and to exhibit comparable behaviour, a deontological approach to collision avoidance can facilitate this.

The literature pertaining to the subject has underscored the challenge of identifying a singular ethical framework that sufficiently addresses the requirements of autonomous vehicles [85]. A mathematical approach to this issue demonstrates the advantages of integrating a deontological perspective in the formulation of ethical guidelines. Specifically, the amalgamation of the imperative to prevent collisions derived from a deontological framework, such as Asimov's laws, with the relative assessment of mobility costs and adherence to traffic regulations establishes a viable reference point for these considerations. Asimov's First Law posits that a robot may not inflict harm upon a human being or, through inaction, permit a human being to suffer harm. Asimov's Second Law asserts that a robot must follow orders issued by humans, except in instances where such orders would conflict with the First Law. It is essential to note that Asimov's Second Law, which mandates compliance with human directives, cannot supersede the First Law; therefore, the imperative to safeguard human life takes precedence over the importance ascribed to human orders.

In order to endow an artificial agent with legal subjectivity, it is imperative that it be endowed with the qualities characteristic of humans, which are designated as the 'central metaphysical functions' [57]: consciousness, intention, and free will.

The concept of consciousness lacks a definitive definition. Furthermore, the evaluation of consciousness is rendered infeasible due to the absence of access to another individual's mind. Nevertheless, it can be inferred based on behavioural states and their analysis. An artificial agent, such as a robot, has the capacity to simulate consciousness by observing its surroundings and incorporating these observations into the learning process of the artificial agent [71].

Aristotle in his work Nicomachean Ethics [90], posited that the notion of responsibility necessitates the fulfilment of at least two conditions: a control condition, which stipulates that the actor must exert control over his or her actions, and an epistemic condition, wherein the individual must possess an understanding of what he or she is doing and the context in which these actions occur. The concept of controllability is intricately connected to moral responsibility through the control principle, which asserts that individuals may only be deemed morally responsible for actions within their realm of control [19, 38]. This principle serves as a widely accepted criterion for the evaluation of morally reprehensible behaviour. Within the scope of ethical analysis, Hevelke and Nida-Rümelin [32] contended that it is unjust to attribute blame to the operator of an autonomous vehicle, particularly those equipped with Automation Levels 4 and 5, for the death of another individual resulting from the actions of said vehicle, especially when the human driver lacked a genuine opportunity to intervene. They stressed that the driver bears no obligation to attend to the road and traffic conditions, nor holds the capacity to intervene when such action might prevent a collision. Nonetheless, it is important to note that philosophers have similarly addressed the concept of 'moral luck' in particular scenarios, wherein an agent may be subjected to moral scrutiny despite the fact that significant elements of their actions are influenced by factors beyond their control [58].

Calo [11] articulated a critical perspective within the scholarly literature concerning the potential for attributing legal subjectivity to artificial agents. This author asserts that an artificial agent lacks the capacity to discern right from wrong, as its operations are fundamentally reduced to adhering to a defined protocol [13] and acting according to directives established by a natural person, encompassing a programmer or user [10].

Himma [33] bases the evaluation of consciousness occurrence in an artificial agent on a subjective assessment made by the evaluator, utilising an analogy to his or her own evaluation of consciousness. Simultaneously, he expresses an agnostic stance regarding the evaluation of consciousness occurrence, while maintaining that the occurrence of consciousness in an artificial agent is not precluded. It should be assumed that an artificial agent, being "aware" of the decisions made, may infer that a human will seek to constrain its pursuit of rewards and will desire to safeguard against this eventuality by developing various types of security software.

In Polish law, the autonomous management system, which encompasses both material objects and intellectual property rights, does not possess a separate legal entity. The recognition of an artificial agent as an independent legal entity to designate a responsible party is currently unattainable, as the artificial agent lacks:

(1) awareness of the decisions made and their consequences,
(2) understanding of the implications of sanctions imposed due to violations of the law related to the decisions made,
(3) the capacity to comprehend the rationale for compensation for damages or redress for wrongs inflicted by the actions of the artificial agent,
(4) the capacity to understand the rationale behind the penalty imposed in relation to the commission of a prohibited act,
(5) the capacity to understand and exercise the rights afforded to the perpetrator (human) within the legal system.

The current legal status of an artificial agent, defined as a robot utilising the capabilities of artificial intelligence as a putative moral agent or an explicit moral agent, is provisional. In functional terms, the following factors influence the modification of this status:

(1) the advancement of engineering technology in the domain of developing self-learning software, which facilitates the creation of tools for improved integration of the artificial agent with its surrounding environment and the human occupants within that environment;
(2) a transformation in the environment in which the artificial agent operates, which would enhance the artificial agent's capacity to exhibit characteristics such as consciousness, free will, and the intention to pursue its own goals.

6.4 The Legal Status of Swarm Intelligence in the Context of Road Transportation

The fifth category of digital mobility services, as delineated in Chap. 4, pertains to the delivery of digital mobility services related to the management of vehicle groups manoeuvring in formations in a coordinated manner. The coordinated movement of vehicles may manifest in various forms, contingent upon the degree of driving automation. With respect to semi-autonomous vehicles, particularly trucks, we can identify platooning as a notable strategy. The term 'platooning' designates a driving approach wherein two or more vehicles follow each other in close proximity. This strategy allows convoy members, in addition to the lead vehicle, to align closely with the vehicle preceding them, thereby enhancing fuel efficiency. Furthermore, convoying diminishes the roadway space occupied by participating vehicles, thus augmenting the effective capacity of the road network. The process of convoying is facilitated by the formation (joining) of multiple vehicles into a cohesive chain (convoy) or by the incorporation of advanced intruder management techniques employing Computer Vision, radar, and lidar sensors. It may also be achieved through the implementation of a GPS-based convoy routing system [53].

Coordinated traffic movement refers to a scenario in which a closed group of vehicles interacts, with the behaviour of one vehicle influencing the actions of the others. The interaction among vehicles within a group may provide enhanced opportunities for effectively executing their assigned tasks, particularly when these vehicles operate autonomously and function as part of an intelligent swarm. In such instances, the increased 'flexibility' in utilising this fleet of vehicles represents added value derived from their integration within a digital mobility service. Consequently, an intelligent swarm is capable of undertaking transportation and logistics tasks that a solitary autonomous vehicle would be unable to accomplish.

It is imperative to first elucidate the mechanisms of swarm intelligence within the domain of autonomous vehicles, followed by an examination of the application of swarming in the context of digital mobility services. This concept should be regarded as an intelligent paradigm that has been effectively employed to address optimisation challenges. As noted by Dagdua et al. [15], swarm intelligence is predicated upon the collective behaviour of various species (for instance, ants, bees, and birds) competing for resources. The primary characteristics of swarm intelligence algorithms include their simplicity and molecular aspect. These algorithms are predicated on agents, such as insects or individuals of a swarm, that exhibit relatively straightforward behaviours and cooperate, manoeuvring within a decision space to accomplish tasks essential for their survival. Swarm intelligence is utilised to tackle issues associated with the optimisation of the transportation of goods and services.

Groups of autonomous collaborative robots may be employed to synergistically execute complex tasks by combining forces and sharing information, or to accomplish designated tasks in a distributed manner, thereby achieving enhanced efficiency through parallelism. Unlike a centralised system, a swarm robotics system lacks a single point of failure, and each component is interchangeable.

Gan et al. [23] elucidated the characteristics of a swarm of autonomous robotic vehicles in their research. They indicated that swarm robotics technology pertains to the capability of robots to operate independently of external infrastructure or centralised control systems. Within a robot swarm, the collective behaviours exhibited by the robots are influenced by their interactions with both the environment and one another. The configuration of the robotic assembly is grounded in the principles of swarm intelligence.

Zhou et al. [89] contend that numerous systems within the transportation and logistics sector should be designed using swarm robotics, including autonomous vehicle systems, autonomous aircraft cargo systems, and automated guided vehicle warehouse systems.

Gan et al. [24] postulated that a collective of truck drivers exhibits certain characteristics akin to swarm intelligence, executing distributed and stochastic road transport services without extensive long-term planning or comprehensive cargo information. Swarm-like behaviours may also be observed in the actions of humans or human-operated vehicles (e.g., trucks) regarding the performance of distributed tasks. Systems of swarm robotics are not limited to merely replicating natural phenomena. The integration of autonomous vehicles to undertake tasks that were previously the responsibility of humans necessitates considering these autonomous vehicles within the framework of autonomous swarm vehicle technology. To achieve this, it is vital to understand the role that humans play.

The subsequent outlined steps executed by an individual driver (truck owner) exemplify the distributed decision-making process:

(1) Identify the most appropriate type of cargo for the specific truck type;
(2) Determine the nearest starting point for the cargo;
(3) Ascertain whether the destination points are known and anticipated;
(4) Contact the cargo owner to verify if the transportation task remains preferable;
(5) If affirmative, negotiate the freight fee;
(6) If the freight fee offered by the cargo owner is deemed acceptable, independently devise the route;
(7) If the cargo is reassigned to other truck drivers or the freight fee is considered unacceptable, reach out to the subsequent cargo owner.

Various individual truck drivers independently engage in the aforementioned decision-making process until all regional transportation assignments are correctly matched with their respective trucks. Upon completing each task, distributed communication technology is utilised within a dynamic task allocation environment. Should a transportation company operate with the same origin and destination, the individual trucks will collectively organise their traffic on the roadway.

Trianni and Campo Alexandre [81] identify the following characteristics of a swarm system: parallelism, scalability, robustness, flexibility, and adaptability to unknown conditions. Contrary to other methodologies that employ multiple robots concurrently, swarm robotics emphasises elements such as the decentralisation of control, limited individual capabilities, absence of global knowledge, and scalability to large groups. Additional attributes may also be noted, including aggregation, patterning, division of labour, clustering, and swarm synchronisation, alongside the assessment of the behavioural characteristics of the swarm of truck drivers [23]. Furthermore, they delineate the following features of swarm intelligence pertinent to the collective behaviours observed in swarm robots, which encompass group aggregation behaviours, member synchronisation and movement coordination within the group, exploratory execution as a collective, and decision-making processes.

The text encompasses all the properties and complexities that make swarm robotics solutions especially suitable. Characteristics such as parallelism, scalability, robustness, flexibility, and adaptability to uncertain conditions are essential for systems faced with such scenarios, and these attributes are exactly what is sought in swarm robotics research. When developing models that capture swarm intelligence, particularly in the transportation sector, the primary focus is on the behaviour of the swarm. To this end, various machine learning models can be employed to predict the upcoming locations of truck drivers. As noted by Gan et al. [24], common features and routines are inherent in the disordered behaviour of truck drivers; the swarm exhibits significant consistency and regularity.

Leminen et al. [48] identified business models for transportation based on swarm intelligence. They propose that vehicles exchange data with neighbouring vehicles and road infrastructure while analysing lane changes, thereby emulating a swarm of fish. Autonomous swarm solutions refer to self-driving vehicles that can function under all weather conditions, across various surface types, and globally. Such systems necessitate the implementation of advanced artificial intelligence, an end-to-end latency of less than one millisecond, and lidar systems designed to operate in challenging weather conditions such as heavy rain or snow. An illustrative example is 'LoadRunner', which is founded on a collaborative logistics system utilising fast automated guided vehicles (AGVs). These autonomously operating elements are capable of negotiating and fulfilling orders individually while also collaborating with one another. This capability facilitates the transportation of loads of varying dimensions and weights. Three representative use cases, specifically parcel sorting, production deliveries, and airport baggage handling, are outlined for LoadRunner [34].

According to Leminen et al. [48], the digital mobility service is based on the developed transformation process of fully autonomous vehicles. Such digital services fall within the business model of flexibility as a service. Digital mobility services operate using swarm intelligence, which provides access to services for recipients, generating value and showcasing characteristics associated with the application of swarm technology to transportation. Autonomous swarm solutions function independently to transport passengers or cargo in open spaces or along designated routes, thereby enabling autonomous (open) transport services. Human intervention or preplanned routes are unnecessary, and vehicles are organised into various arrangements to fulfil their tasks in autonomous swarm solutions. Flexibility as a service expands the notion of agile conceptualisation and commercialisation, allowing for autonomous reconfiguration (absent formal decisions from ecosystem partners) of commercialised concepts within the framework of the autonomous vehicle service ecosystem.

Flexibility as a Service advocates for continuous, hyper-fast, and iterative reconfigurations of digital servitisation business models, particularly in the context of autonomous vehicles. In this scenario, autonomous vehicles within a swarm determine their organisation autonomously. The flexibility service—which facilitates business model fluidity—addresses the significant expectations outlined in its definition. Business model fluidity is not static; rather, it encompasses rapid changes and reconfigurations that the swarm organises and executes automatically. In extreme cases, each customer and task may necessitate a distinct business model. The swarm autonomously organises, reconfigures, and optimises both itself and the business model based on the active customer base and the specific tasks to be accomplished [48].

The explanations above enable a thorough discussion about the legal status of a group of autonomous vehicles within the context of the implemented digital mobility services.

At the outset, it is crucial to recognise that there is no established legal definition of a swarm in the context of autonomous vehicles. The legislator has not directly addressed the determination of the legal status of a swarm of autonomous vehicles. In evaluating the legal status of an intelligent swarm, it is essential to consider this matter concerning a closed group of artificial agents, namely robots equipped with software that enables them to exhibit various attributes that facilitate the creation of a digital mobile service and the resulting value for both the customer and the supplier. Categorising a swarm as a closed assembly of robots entails the presumption that each robot possesses characteristics of collective behaviour, including group aggregation, member synchronisation, and coordination of movements within the group, along with collaborative exploration, execution, and decision-making. Additional attributes of a swarm include parallelism, scalability, robustness, flexibility, and the ability to adapt to unforeseen conditions. Robots operate within a decentralised system, carrying out their designated tasks while adjusting the execution processes based on the behaviours of other group members.

The reference point for utilising a swarm of intelligent autonomous vehicles is the execution of intricate tasks, which has necessitated the formulation of a digital mobility service. Accordingly, the objective of providing a digital mobility service based on an intelligent swarm of autonomous vehicles is to tackle complex issues related to transportation and logistics.

In Article 3(1) of Regulation (EU) 2024/1689 ('AI Act'), the legislator outlines the criteria for the establishment of an 'artificial intelligence system', based on four key elements. Specifically, such a system:

(1) Receives input from a machine and/or a human;
(2) Infers methods to attain a specified set of human-defined objectives through the processes of learning, reasoning, or modelling; and
(3) Produces output in the following formats: content, predictions, recommendations, or decisions that influence the environments with which it engages.

By incorporating the aforementioned definition into the AI Act, the European Union legislator has determined, at least at this juncture, that artificial intelligence lacks legal personality, a principle that should be extended to autonomous vehicles operating at the fourth and fifth levels of driving automation. The AI Act should encompass individual robots that form a swarm. Nevertheless, there are currently no regulations addressing an 'intelligent swarm' when analysed from a functional standpoint, which considers the distinct relationships among the swarm's members concerning the features of the aforementioned implementation. The European Union legislator in the AI Act does not recognise the interrelationship among the robots in the group regarding the fulfilment of certain complex fundamental purposes for which the intelligent swarm is deployed. Had the AI Act included a definition of "intelligent swarm," the legislator could concurrently determine the requirements for applying the intelligent swarm without regard to the application of the AI Act to the individual robots comprising it. Consequently, this would simplify the requirements of the AI Act as they pertain to the swarm.

If forthcoming legal regulations pertain to the legal status of artificial intelligence, they should also encompass every autonomous vehicle within an intelligent swarm. In this context, the regulations may specifically target the entire collective of autonomous vehicles that are part of an intelligent swarm, considering the implementation of shared complex objectives that underpin digital mobility services. The legislator provides for regulations applicable to entities that are both legally and factually interrelated, including capital considerations. For instance, Polish law encompasses regulations relevant to a capital group. The Polish legal framework presents various definitions and interpretations of a capital group, which are reflected, among others, in the Accounting Act [4]. According to this Act, a capital group possesses distinct characteristics and specific obligations. Additionally, the concept of a tax capital group is delineated in the Corporate Income Tax Act [76].

The definition of a capital group is further elucidated in the Competition and Consumer Act [1]. Article 4(14) of this Act delineates 'capital group' as encompassing all entrepreneurs, both directly and indirectly controlled by a single entrepreneur, including that entrepreneur. This definition applies to natural persons, legal entities, and organisational units lacking legal personality, which are granted by a separate act, the legal capacity to independently operate a business. The aforementioned definitions may guide further investigation into the legal regulation of intelligent swarms. A key aspect of this inquiry is the necessity for members of an intelligent swarm to be accorded legal entity status. Moreover, these definitions also contemplate capital and personal relationships among group members, particularly in the composition of legal persons' governing bodies, which are not pertinent to intelligent swarms. A defining characteristic of the 'capital group', as articulated in the Competition and Consumer Act (2007), is the control one entity exercises over the other members of the group. However, within the dynamics of intelligent swarms, such control, as described in the definitions, does not manifest. The execution of tasks within this context is decentralised. Importantly, when examining prospective legal frameworks for intelligent swarms, the civil partnership structure detailed in Title XXXI of the Polish Civil Code [77] is significant. Article 860 § 1 states that the essence of a civil partnership involves its members striving towards a common economic objective, which would be unattainable if each partner operated independently. Similar collaborative motivations can be discerned in intelligent swarms, despite the fact that the artificial agents involved do not possess distinct legal subjectivity.

6.5 Conclusions

Digital mobility services may influence drivers in several capacities. These services may:

(1) exert a direct influence on the driver's performance of tasks associated with vehicle operation;
(2) indirectly impact the driver's execution of activities related to driving the vehicle;
(3) neither directly nor indirectly affect vehicle operation, but pertain to the organisation of transportation, thereby facilitating the driver's engagement in tasks associated with operating an autonomous vehicle.

Consequently, the provision of these services is contingent upon the legal definition of the status of the 'driver' as the direct or indirect beneficiary.

The concept of a 'driver' possesses a normative significance. This notion should be examined within two legal frameworks. In the context of administrative law, the term 'driver' specifically refers to a natural person who, having fulfilled the requisite preliminary conditions established by law, such as attaining the appropriate age and possessing a suitable level of health, has successfully completed both a theoretical and practical examination pertaining to the operation of a vehicle on public roads and has subsequently been granted a permit for motor vehicle operation. By the issuance of such a permit, public authorities affirm that the individual has satisfied the necessary prerequisites for obtaining the designation of a driver. Furthermore, the law distinguishes the status of an 'actual driver' as an individual who is actively operating a vehicle, irrespective of whether they possess the requisite authorisation at that moment to do so.

The legislator imposes several legal requirements upon the driver:

(1) to secure the right to operate vehicles,
(2) to meet obligations towards administrative bodies, road traffic control authorities, passengers, and other road users pertaining to the operation of a vehicle in road traffic,
(3) to meet obligations concerning the admission of the vehicle into traffic, the fulfilment of which is verified through the vehicle registration certificate.

The nature of the digital mobility services provided also influences the fulfilment of these obligations. Civil, criminal, and administrative liability is asserted against the actual driver.

The definitions of 'actual driver' and 'driver' pertain to the definition of 'vehicle'. In the context of autonomous vehicle operation, particularly regarding driving automation modes that exclude human involvement in the steering function, it is thus imperative to consider the legal status of an autonomous vehicle operating autonomously, as well as within interacting groups of vehicles (swarm intelligence).

The Convention on Road Traffic [14] requires that each vehicle have a 'real driver', although this does not necessarily have to be a human. Under Polish law, recognising an artificial entity as a separate legal person to designate the responsible entity is currently not feasible, as the artificial entity lacks: (1) awareness of the decisions made and their consequences, (2) understanding of the significance of sanctions imposed as a result of legal breaches in relation to the decisions made, (3) the ability to comprehend the grounds for compensation for harm or damage caused by its behaviour, (4) the capacity to understand the rationale behind the punishment imposed for committing a prohibited act, and (5) the ability to understand and exercise the rights that human perpetrators are entitled to under the legal system.

Provided that the human driver is the recipient of digital mobility services, their delivery is constrained by the human capacity to process information in a manner that does not endanger road safety. With regard to the use of autonomous vehicles, designating a human as the recipient of digital mobility services restricts these services to the third level of automation.

References

1. Act of 16 February 2007 on Competition and Consumer Protection (Consolidated Text: Journal of Laws of 2024, Item 1616)
2. Act of 20 June 1997—Road Traffic Law (Consolidated Text: Journal of Laws of 2024, Item 1251)
3. Act of 21 March 1985 on Public Roads (Consolidated Text: Journal of Laws of 2024, Item 320, as Amended)
4. Act of 29 September 1994 on Accounting (Consolidated Text: Journal of Laws of 2023, Item 120, as Amended)
5. Altunyaldiz, Z.: Legal aspects of "autonomous" vehicles. Committee on Legal Affairs and Human Rights. Draft resolution and draft recommendation unanimously adopted by the committee, 9 Sept 2020. http://www.assembly.coe.int/LifeRay/JUR/Pdf/DocsAndDecs/2020/AS-JUR-2020-20-EN.pdf. Accessed 21 Feb 2025
6. Asimov, I.: Robot. London (1950)
7. Bertolini, A.: Robots as products: the case for a realistic analysis of robotic applications and liability rules. Law Innov. Technol. **5**(2) (2013)
8. Brodsky, W.: The effects of music tempo on simulated driving performance and vehicular control. Transp. Res. Part F **4**, 219–241 (2002)
9. Buchała, K.: Przestępstwa i wykroczenia przeciwko bezpieczeństwu w komunikacji drogowej, pp. 124–125. Komentarz, Bydgoszcz (1997)
10. Calo, R.: Robotics and the lessons of cyberlaw. Calif. Law Rev. **103** (2015)
11. Calo, R.: Commuting to Mars: a response to professors Abraham and Rabin. Virginia Law Rev. **105**, 84 (2019)
12. Carey, N., Lienert, P.: Insight: truly autonomous cars may be impossible without helpful human touch. Reuters, 19 Sept 2022. https://www.reuters.com/technology/truly-autonomous-cars-may-be-impossible-without-helpful-human-touch-2022-09-12/. Accessed 21 Feb 2025

13. Chustecki, J., et al.: Vademecum teleinformatyka. Warszawa (1999)
14. Convention on Road Traffic, Drawn Up in Vienna on 8 November 1968 (Journal of Laws 1988, No. 5, Item 40, as Amended). https://archives.ungeneva.org/un-conference-on-road-traffic-vienna-1968-final-text-of-convention-on-road-traffic-1968-and-of-the-final-act-of-the-conference. Accessed 5 Feb 2025
15. Dagdua, Z.C., Mirchev, M.: When evolutionary computing meets astro- and geoinformatics, chap. 15. In: Škoda, P., Fathalrahman, A. (eds.) Knowledge Discovery in Big Data from Astronomy and Earth Observation. Astrogeoinformatics (2020)
16. Dubber, M.D., Pasquale, F., Das, S. (eds.): The Oxford Handbook of Ethics of AI. Oxford University Press, Oxford, UK (2020)
17. Duffy, S., Hopkins, J.: Sit, stay, drive: the future of autonomous car liability. SMU Sci. Technol. Law Rev. **16**(3) (2013)
18. Echeto, J., Santos, M., Romana, M.G.: Automated vehicles in swarm configuration: simulation and analysis. Neurocomputing **501**(28), 679–693 (2022)
19. Fischer, J.M., Ravizza, M.: Responsibility and Control: A Theory of Moral Responsibility. Cambridge (1998)
20. Fleetwood, J.: Public health, ethics, and autonomous vehicles. Am. J. Public Health **107**(4) (2017)
21. Floridi, L., Sanders, J.W.: On the morality of artificial agents. Minds Mach. **14**(3) (2004)
22. Funkhouser, K.: Paving the road ahead: autonomous vehicles, products liability, and the need for a new approach. Utah Law Rev. **1** (2013)
23. Gan, M., Marco, Y., Liu, X., Zhu, D.: Whereabouts of truckers: an empirical study of predictability, pp. 1–14 (2019)
24. Gan, M., Qian, Q., Li, D., Ai, Y., Liu, X.: Capturing the swarm intelligence in truckers: the foundation analysis for future swarm robotics in road freight. Swarm Evol. Comput. **62**, 100845 (2021). https://doi.org/10.1016/j.swevo.2021.100845
25. Garza, A.: Look ma, no hands!: wrinkles and wrecks in the age of autonomous vehicles. New Engl. Law Rev. **46**(3) (2012)
26. Gasser, T.M.: Grundlegende und spezielle Rechtsfragen für autonome Fahrzeuge. In: Maurer, M., Gerdes, J.C., Lenz, B., Winner, H. (eds.) Autonomes Fahren, Technische, rechtliche und gesellschaftliche Aspekte, pp. 543–544. Berlin (2015)
27. Gerdes, J.C., Thornton, S.M.: Implementable ethics for autonomous vehicles. In: Maurer, M., Gerdes, J.C., Lenz, B., Winner, H. (eds.) Autonomes Fahren. Technische, rechtliche und gesellschaftliche Aspekte. Berlin (2015)
28. Goodall, N.J.: Machine ethics and automated vehicles. In: Meyer, G., Beiker, S. (eds.) Road Vehicle Automation. London (2014)
29. Guizzo, E.: How Google's self-driving car works. Google engineers explain the technology behind their self-driving car and show videos of road tests. IEEE Spectrum, 18 Oct 2011. https://spectrum.ieee.org/how-google-self-driving-car-works. Accessed 21 Feb 2025
30. Gurney, J.: Sue my car not me: products liability and accidents involving autonomous vehicles. Univ. Illinois J. Law Technol. Policy **2** (2013)
31. Haigney, D.E., Taylor, R.G., Westerman, S.J.: Concurrent mobile (cellular) phone use and driving performance: task demand characteristics and compensatory processes. Transp. Res. Part F **3**, 113–121 (2000)
32. Hevelke, A., Nida-Rümelin, J.: Responsibility for crashes of autonomous vehicles: an ethical analysis. Sci. Eng. Ethics **21**(3) (2015)
33. Himma, K.E.: Artificial agency, consciousness, and the criteria for moral agency: what properties must an artificial agent have to be a moral agent? Ethics Inf. Technol. **11** (2009)
34. Hompel, M., Bayhan, H., Behling, J., Benkenstein, L., Emmerich, J., Follert, G., Grzenia, M., Hammermeister, C., Hasse, H., Hoening, D., Hoppe, C., Kerner, S., Klokowski, P., Korth, B., Kuhlmann, G., Leveling, J., Lünsch, D., Mendel, R., Menebroker, F., Nettstrater, A., Pieperbeck, J., Prasse, C., Roeder, S., Roidl, M., Rotgeri, M., Sparer, D., Walter, S., Wolf, O.: Technical report: LoadRunner, a new platform approach on collaborative logistics services. Logist. J. **2020**, 1–14 (2020). https://www.logistics-journal.de/not-reviewed/2020/10/5114. Accessed 1 Feb 2025

35. Horberry, T., Anderson, J., Regan, M.A., Triggs, T.J., Brown, J.: Driver distraction: the effects of concurrent in-vehicle tasks, road environment complexity and age on driving performance. Accid. Anal. Prev. **38**, 185–191 (2005)
36. Imai, T.: Legal regulation of autonomous driving technology: current conditions and issues in Japan. IATSS Res. **43** (2019)
37. Ingrassia, P.: Look, no hands! Test driving a Google car. Reuters, 17 Aug 2014, https://www.reuters.com/article/technology/look-no-hands-test-driving-a-google-car-idUSKBN0GH02P/. Accessed 21 Feb 2025
38. Johnson, D.G.: Technology with no human responsibility? J. Bus. Ethics **127**(4), 707–708 (2015)
39. Judgment of the Supreme Court of 20 October 2022, IV KK 341/21, LEX No. 3448198
40. Judgment of the Supreme Court of 29 September 1993, III KRN 127/93, LEX No. 22087
41. Jurczak, M., Pawlicka, K.: Transport autonomiczny jako element rozwoju miasta zrównoważonego. Stud. Miejskie **45** (2023)
42. Kapitan, T.: The free will problem. In: Audi, R. (ed.) Cambridge Dictionary of Philosophy, p. 326. Cambridge (1999)
43. Keng, S., Wang, W.: Artificial intelligence (AI) ethics: ethics of AI and ethical AI. J. Database Manage. **31**(2), 74–86 (2020)
44. Kitano, N.: 'Rinri': an incitement towards the existence of robots in Japanese society. Int. Rev. Inf. Ethics **6**, 78–83 (2006)
45. Kochanowski, G.: Przestępstwa i wykroczenia drogowe. Komentarz, Warszawa (1991)
46. Krügel, S., Uhl, M.: Autonomous vehicles and moral judgments under risk. Transp. Res. Part A Policy Pract. **155**, 1–10 (2022)
47. Księżak, P., Wojtczak, S.: Toward a Conceptual Network for the Private Law of Artificial Intelligence. Springer (2024)
48. Leminen, S., Rajahonka, M., Wendelin, R., Westerlund, M., Nystrom, A.: Autonomous vehicle solutions and their digital servitization business models. Technol. Forecast. Soc. Change **185**, Article 122070 (2022). https://doi.org/10.1016/j.techfore.2022.122070
49. Levinas, E.: Levinas on the Difference Between Morality and Ethics—Lecture 7 (1991). https://drwilliamlarge.wordpress.com/2016/08/29/levinas-on-the-difference-between-morality-and-ethics-lecture-7/. Accessed 23 Apr 2024
50. Lin, P., Bekey, G., Abney, K.: Autonomous Military Robotics: Risk, Ethics, and Design. Raport DLA US Department of Navy, Office of Naval Research (2008). https://digitalcommons.calpoly.edu/cgi/viewcontent.cgi?article=1001&context=phil_fac. Accessed 21 Feb 2025
51. Liu, P., Du, M., Li, T.: Psychological consequences of legal responsibility misattribution associated with automated vehicles. Ethics Inf. Technol. **23**(4) (2021)
52. Mackie, T.: Proving liability for highly and fully automated vehicle accidents in Australia. Comput. Law Secur. Rev. **3** (2018)
53. Malao, A., Pawar, A., Bhamre, H., Tagare, P., Sorate, G.: Vehicle platoon formation and management. Int. Res. J. Eng. Technol. (IRJET) **08**(05) (2021). ISSN: 2395-0056
54. Mamak, K., Glanc, J.: Problems with the prospective connected autonomous vehicles regulation: finding a fair balance versus the instinct for self-preservation. Technol. Soc. **71** (2022)
55. Matsuzaki, H.: Gehorsamer Diener oder gleichberechtigter Partner? Überlegungen zum gesellschaftlichen Status von humanoiden Robotern in Japan. Technikgeschichte **77**(4) (2010)
56. Mezgár, I., Váncza, J.: From ethics to standards—a path via responsible AI to cyber-physical production systems. Annu. Rev. Control **53**, 391–404 (2022)
57. Moor, J.H.: Four kinds of ethical robots. Philosophy Now (2009). https://philosophynow.org/issues/72/Four_Kinds_of_Ethical_Robots. Accessed 21 Feb 2025
58. Nagel, T.: Mortal Questions. New York (1979)
59. Neudert, L.-M., Knuutila, A., Howard, P.N.: Global attitudes towards AI, machine learning & automated decision making. Working paper 2020.10. Oxford Commission on AI & Good Governance (2020)

60. Pattinson, J.A., Chen, H., Basu, S.: Legal issues in automated vehicles: critically considering the potential role of consent and interactive digital interfaces. Humanit. Soc. Sci. Commun. **7**(1) (2020)
61. Pawelec, K.J.: Wypadek drogowy. Warszawa (1999)
62. Regulation (EU) 2024/1689 of the European Parliament and of the Council of 13 June 2024 Laying Down Harmonised Rules on Artificial Intelligence and Amending Regulations (EC) No 300/2008, (EU) No 167/2013, (EU) No 168/2013, (EU) 2018/858, (EU) 2018/1139 and (EU) 2019/2144 and Directives 2014/90/EU, (EU) 2016/797 and (EU) 2020/1828 (Artificial Intelligence Act) (Text with EEA Relevance), E/24/2024/REV/1, OJ L, 2024/1689, 12.7.2024, ELI: http://data.europa.eu/eli/reg/2024/1689/oj. Accessed 21 Feb 2025
63. Regulation of the Prime Minister of 24 November 2003 on the Amount of Fines Imposed in the Form of Penalty Tickets for Selected Types of Petty Offences (Consolidated Text: Journal of Laws of 2013, Item 1624, as Amended)
64. Resolution of the Supreme Court (CI) of 28.02.1975, V KZP 2/74, OSNKW 1975, No. 3–4, Item 33
65. Resolution of the Supreme Court of 8 January 2019, II KK 145/18, LEX No. 2618470
66. Robaczyński, W.: Odpowiedzialność za szkody wyrządzone przez pojazdy autonomiczne. Forum Prawnicze **1**(69) (2022)
67. Rousseau, D.M., Sitkin, S.B., Burt, R., Camerer, C.: Not so different after all: a cross-disciplinary view of trust. Acad. Manage. Rev. **23**(3) (1998)
68. Seiniger, P., Bartels, O., Pastor, C., Wisch, M.: An open simulation approach to identify chances and limitations for vulnerable road user (VRU) active safety. Traffic Inj. Prev. **14** (2013)
69. Shenzhen gives green light to fully autonomous vehicles, 8.07.2022. https://english.www.gov.cn/news/topnews/202207/08/content_WS62c77b57c6d02e533532d5cc.html#:~:text=Shenzhen%2C%20a%20city%20that%20pioneered%20reform%20and%20opening-up,first%20regulation%20tailored%20for%20smart%20and%20internet-connected%20vehicles. Accessed 23 Feb 2025
70. Sheridan, T.B., Parasuraman, R.: Human-automation interaction. Rev. Hum. Factors Ergon. **1** (2005)
71. Solum, L.M.: Legal personhood for artificial intelligences. North Carolina Law Rev. **70**(4) (1992)
72. Stefański R.A.: Przestępstwa drogowe w nowym kodeksie karnym. Kraków (1999)
73. Sturma, D.: Autonomie: Über Personen Künstliche Intelligenz und Robotik. In: Christaller, T., Wehner, J. (eds.) Autonome Maschinen. Westdeutscher Verlag (2003)
74. Supreme Court Judgment of 16 July 1975, I KRN 79/75, Not Published
75. Supreme Court Judgment of 19 February 1974, Rw 53/74, OSNKW 1974, No. 6, Item 125
76. The Act of 15 February 1992 on Corporate Income Tax (Consolidated Text: Journal of Laws of 2025, Item 278, as Amended)
77. The Act of 23 April 1964—The Civil Code (Consolidated Text: Journal of Laws of 2024, Item 1061, as Amended)
78. The Act of 24 August 2001—the Code of Procedure in Petty Offences (Consolidated Text: Journal of Laws of 2024, Item 977, as Amended)
79. The Act of 6 June 1997—The Penal Code (Consolidated Text: Journal of Laws of 2025, Item 383)
80. The Constitution of the Republic of Poland of 2 April 1997 (Journal of Laws No. 78, Item 483, as Amended)
81. Trianni, V.: Campo Alexandre, A.: Fundamental collective behaviors in swarm robotics. In: Springer Handbook of Computational Intelligence (2015)
82. Ünal, A.B., Steg, L., Epstude, K.: The influence of music on mental effort and driving performance. Accid. Anal. Prev. **48**, 271–278 (2012)
83. van den Hoven van Genderen, R.: Do we need new legal personhood in the age of robots and AI? In: Corrales, M., Fenwick, M., Forgó, N. (eds.) Robotics, AI and the Future of Law. Springer Nature Singapore Pte Ltd (2018)

84. Vladeck, D.: Machines without principals: liability rules and artificial intelligence. Washington Law Rev. **89**(1) (2014)
85. Wallach, W., Allen, C.: Moral Machines: Teaching Robots Right from Wrong. Oxford (2009)
86. Wiseman, Y., Grinberg, I.: The trolley problem version of autonomous vehicles. The Open Transp. J. **12**, 105–113 (2018). https://doi.org/10.2174/18744478018120100105
87. Zech, H.: Zivilrechtliche Haftung für den Einsatz von Robotern—Zuweisung von Automatisierungs- und Autonomierisiken. In: Gless, S., Seelmann, K. (eds.) Intelligente Agenten und das Recht. Baden-Baden (2016)
88. Zhai, S., Wang, L., Liu, P.: Not in control, but liable? Attributing human responsibility for fully automated vehicle accidents. Engineering **33**, 121–122 (2024)
89. Zhou, S., Phielipp, M.J., Sefair, J.A., Walker, S.I., Ben Amor, H.: Clone swarms: learning to predict and control multi-robot systems by imitation. In: Proceedings of the IEEE International Conference on Intelligent Robots and Systems (2019)
90. Aristotle, Bartlett R.C., Collins S.D.: Aristotle's Nicomachean Ethics, University of Chicago Press, Chicago pp. 339, (2011) ISBN: 9780226026749, 9780226026756, 0226026744, 0226026752.

Chapter 7
Challenges of Privacy Protection in the Context of Providing Digital Mobility Services

7.1 Sources of Data Protection Regulations

Pagani and Pardo [74] assert that digitalisation facilitates the dissolution of boundaries between stakeholders, fosters partnerships among entities, and permits both upstream and downstream participants to engage directly. Furthermore, Theodorou et al. [108] indicate that artificial intelligence has the capacity to deduce patterns and potentially derive conclusions from data; as of now, AI technologies frequently rely on machine learning and/or neural network-based paradigms.

The processing of customer data substantiates the diverse range of digital mobility services outlined in Chap. 4. The use of digital tools based on big data analytics, artificial intelligence capabilities, and B2B buyer–seller interactions reliant on reputational assessments necessitates processing substantial volumes of data. The ability to manage data and transform it into knowledge becomes a critical component in interactions with customers and sales [98].

In designing and analysing a business model, it is imperative to consider the dynamic interrelationships among various stakeholders, technological advancements, and market dynamics. In light of these novel circumstances, there is a lack of a comprehensive theoretical and legal framework; therefore, applying contingency theory is deemed necessary.

Contingency theory posits that organisational performance is determined by the alignment of the organisation's characteristics with the contingencies that reflect its situational context [22]. While contingency theory asserts that elements become interconnected by virtue of their occurrence contingencies, independent of their content, similarity theory maintains that such elements become associated due to the similarity of their content, also independent of contingency. Furthermore, the similarity of content necessitates commensurate representations, which are representations whose mutual relationships can be evaluated in terms of representational overlap [92].

Consequently, organisations will endeavour to align their strategies with contingency factors that may emerge in rapidly changing environments through a dynamic

A. Krasuski, *Legal Challenges of a Transformation from a Product-Centric to a Service-Centric Business Model and Procedures*, Studies in Systems, Decision and Control 610, https://doi.org/10.1007/978-3-031-98901-8_7

and ongoing process [98]. The use of digital tools for the development of business models within the mobility sector, aimed at cultivating business relationships with customers, is inherently linked to data processing that also considers the identification data of these individuals. As a result, this gives rise to inquiries regarding the establishment of suitable legal safeguards for these individuals.

The foundational reference for these considerations regarding the example of Japan is the Constitution of Japan, which, in Article 13, asserts that: 'All of the people shall be respected as individuals. Their right to life, liberty, and the pursuit of happiness shall, to the extent that it does not interfere with the public welfare, be the supreme consideration in legislation and in other governmental affairs' [95].

The use of digital tools for developing business models in the mobility sector, which aim to foster customer relationships, involves data processing that also includes the identification data of these individuals. This situation raises concerns about establishing appropriate legal safeguards for them. This is articulated in (EU Convention 108, 1981) [18], which states in its preamble that: 'Considering that it is necessary to secure the human dignity and protection of the human rights and fundamental freedoms of every individual and, given the diversification, intensification and globalisation of data processing and personal data flows, personal autonomy based on a person's right to control of his or her personal data and the processing of such data'.

Although not explicitly articulated in Regulation (EU) 2016/679 ('GDPR') [79], academic literature [32] highlights that the GDPR should be interpreted with due regard for human dignity as the principal tenet. Moreover, prior viewpoints [35] concerning the values embodied within the European privacy framework emphasise the prioritisation of privacy standards based on the intrinsic dignity of individuals.

The national legislation of the Member States adopts a distinct perspective on the right to privacy, which remains aligned with the subjective right to dignity. For instance, under Polish law, Article 47 of the Constitution of the Republic of Poland [96] establishes the right of every individual to legal protection concerning their private life, family life, honour, and reputation, as well as the right to determine matters affecting their personal life. Moreover, Article 51(2) of the Polish Constitution (1997) expressly prohibits public authorities from acquiring, collecting, and disseminating 'other information about citizens than is necessary in a democratic state governed by the rule of law'.

Although the latter provision is directed at the legislator, it is evident, in light of the jurisprudence of the Polish Constitutional Court (Judgment SK 26/16 [51]), that the subjective right of the individual to the protection of his or her informational autonomy—including, in particular, the protection of personal data—should be derived from this judgment (Judgment SK 40/01 [46]; Judgment K 45/02). Informational autonomy is an essential component of the right to privacy and entails the right to determine for oneself whether to disclose information pertaining to one's person to others, as well as the right to exercise control over this information when it is held by others (Judgment K 8/04 [49]; Judgment K 33/08 [50]).

The concepts of 'respect', as articulated in the Japanese Constitution, and 'dignity' within the European legal framework exhibit analogous significances in the relevant literature [68]. Generally, Japan's prevailing privacy legislation, notably the Act on the Protection of Personal Information (Act No. 57, 2003) [1], adopts a human-centric perspective within the privacy sector. This orientation is further evidenced by the Cabinet Office of Japan's dissemination of the Social Principles of Human-centric AI in 2019, which asserts that "[w]e need to construct a society in which human dignity is respected, such as people demonstrating human abilities, showing great creativity, doing challenging work and living a lively life materially and spiritually, by using AI as their instruments" [12].

The obligations regarding the design and operation of artificial intelligence (AI) systems in relation to personal data are delineated in Regulation (EU) 2024/1689 (commonly referred to as the 'AI Act') [86]. Ren and Du [87] elucidate that the AI Act expressly prohibits the utilisation of AI systems for social scoring. This practice evaluates or classifies individuals over time based on their social behaviour or known or predicted personal and personality characteristics. Such methodologies are deeply rooted in significant concerns surrounding privacy, autonomy, and discrimination. Furthermore, social scoring systems have the potential to engender dystopian scenarios by fostering extensive surveillance and profiling predicated on an individual's behaviour and characteristics [73]. This practice infringes upon the right to privacy by persistently monitoring and analysing personal activities [93]. Moreover, these systems can encroach upon individual autonomy, thereby influencing or dictating personal choices and freedoms [90]. The application of personal or personality characteristics for scoring purposes may result in discrimination, as these systems could inadvertently reinforce societal biases and unjustly penalise certain groups based on arbitrary or prejudiced criteria [10]. The AI Act, among other stipulations, prioritises systems wherein social scoring leads to harmful or unjust treatment, thus safeguarding individuals from arbitrary and opaque assessments that may profoundly affect their lives and social interactions. Furthermore, the Act imposes stringent regulations on the use of real-time remote biometric identification systems in public spaces by law enforcement authorities. The employment of AI technologies for the analysis of recorded footage in public contexts is subject to more rigorous controls due to widespread apprehensions regarding mass surveillance and its implications for privacy and civil liberties. While not categorically prohibitive, the Act establishes high standards for the utilisation of such technologies, constraining their application to scenarios involving grave criminal activities or significant threats, such as terrorism, in order to strike a balance between the legitimate requisites of law enforcement and the safeguarding of fundamental rights. The prohibitions enshrined in the AI Act manifest a legislative intent centred on averting AI applications that could potentially undermine human dignity, exploit vulnerabilities, encroach upon privacy, or facilitate unwarranted surveillance. These provisions emerge against the backdrop of swift technological progression, wherein the capacity for AI to influence numerous facets of daily life is considerable.

7.2 Legal Challenges Regarding Data Protection in Relation to the Use of Autonomous Vehicles

7.2.1 The Protection of Human Autonomy and Privacy Concerning the Use of Autonomous Vehicles

The intricate nature of autonomous vehicle operations and their interactions with the surrounding environment necessitates a thorough examination of the legal foundation and nature of data in relation to the exercise of the right to privacy. This encompasses the legal classification of data processed in connection with the movement of an autonomous vehicle, the specific purposes and legal basis for the processing of personal data, the legal status of the responsible entities, and the obligations pertinent to the processing of such data.

The analysis presented in this monograph is based on the premise that effectively implementing the rights and freedoms of natural persons whose personal data are processed in relation to the movement of an autonomous vehicle significantly enhances trust in such vehicles, thereby facilitating a revolutionary transformation in transportation within this domain. The formal-dogmatic method was employed to conduct this analysis.

For further considerations, it is advisable to refer to the concept of an 'autonomous vehicle' as discussed in Chap. 2.

The concept of autonomy possesses a multifaceted significance. It should also be associated with human rights, encompassing the right to privacy. In legal doctrine [103], a distinction is made between positive and negative autonomy. Positive autonomy encompasses an individual's capacity to make decisions regarding themselves, including choices related to interactions with others, such as utilising various modes of transportation. Conversely, negative autonomy pertains to an individual's right to request that third parties impose limitations on access to them. This framework highlights the correlation between negative autonomy and privacy, which is defined as the "right to be left alone" [35].

The Polish Constitutional Court has previously highlighted that the protection of private life, as articulated in Article 47 of the Constitution of the Republic of Poland (Polish Constitution 1997), encompasses, among other aspects, information autonomy. This autonomy signifies the right to independently determine the disclosure of personal information to others and to maintain control over this information when it is held by external entities (refer to Rulings of the Constitutional Court: U 3/01 [45], K 41/02 [47]). The Court concluded that, akin to other constitutional rights and freedoms, the safeguarding of privacy and information autonomy is not absolute, primarily due to the demands of communal living. In this context, the principle of proportionality assumes critical significance: "any (...) interference in privacy must take into account its protection as a constitutionally protected good and the constitutional principles indicating the limits and premises that must be maintained in the event of interference. The obligation to disclose information about oneself, constituting a limitation of information autonomy, may therefore be made only by statute (…) and exclusively within the limits consistent with the constitutional principle of proportionality" (see Judgement of the Constitutional Court: K 41/02).

Article 47 of the Constitution of the Republic of Poland establishes the right of every individual to legal protection of their private and family life, honour, and reputation, as well as the right to make determinations regarding their personal life. Furthermore, Article 51(2) of the Polish Constitution prohibits public authorities from obtaining, collecting, and disseminating 'information about citizens other than necessary in a democratic state of law'. Although this particular provision is directed at the legislator (refer to Judgement of the Constitutional Court: SK 26/16 [44]), it implies the subjective right of individuals to protect their informational autonomy, which encompasses, in particular, the safeguarding of personal data (see Judgement of the Constitutional Court: K 45/02 [48]). Informational autonomy represents a critical element of the right to privacy, embodying the individual's right to independently decide on the disclosure of information concerning themselves to others, as well as the right to exercise control over such information when other entities possess it (see Judgement of the Constitutional Court: K 8/04).

The Constitutional Court defines proportionality as comprising three fundamental principles: the principle of usefulness, the principle of necessity, and the principle of proportionality in its strictest sense, which entails the prohibition of excessive interference (see Judgement of the Constitutional Court: K 15/98 [43]).

In conclusion, an individual's decision to utilise an autonomous vehicle fundamentally reflects the principle of individual autonomy. Conversely, this autonomy is constrained concerning certain social groups. Various subjective factors, such as disability and age, alongside infrastructural obstacles like the lack of alternative public transportation, render the use of an autonomous vehicle a necessity [23]. Consequently, in light of Article 47 of the Constitution of the Republic of Poland, it is imperative for the state to take an active role in facilitating individuals' rights to make decisions regarding their personal lives and safeguarding privacy [89].

7.2.2 Collection of Personal Data in Relation to the Use of an Autonomous Vehicle

The specifics of autonomous vehicles relate to the processing of data. Autonomous vehicles are equipped with multiple sources of data collection regarding the environment in which the vehicle is moving [33]. Cameras and radars are used to detect nearby objects and measure the distance to other vehicles. Additionally, light and distance detection tools can be employed to create a scanned 360° representation of the vehicle's surrounding environment using laser rangefinders (On the machine learning techniques as well as sensors used in the DARPA Grand Challenge in 2005 see Thrun et al. [97]). Subsequently, software algorithms interpret this data and control the vehicle's trajectory [4].

The procedure by which algorithms contribute to the steering of a vehicle can be delineated into three distinct stages, namely:

(1) vehicle localisation,
(2) obstacle detection and behaviour prediction, and
(3) path planning [41].

The Automated Driving System ('ADS') initially determines its geographical position by utilising Global Positioning System ('GPS') technology in conjunction with an inertial navigation system. This system employs gyroscopes and accelerometers to compute the position, orientation, and velocity of the vehicle [4].

The localisation and mapping algorithms determine both the global and local positions of the vehicle while concurrently mapping the environment by utilising sensor data and additional perception outputs [66]. The integration of various sensors and camera modules within an autonomous vehicle results in the processing of substantial quantities of data, upon which decisions regarding the vehicle's steering are formulated. An individual autonomous vehicle has the capability to accumulate over 300 terabytes of data within a single year [99].

In the realm of privacy protection, two theoretical constructs have emerged in the literature regarding the classification of autonomous vehicles: specifically, self-contained autonomous vehicles and interdependent autonomous vehicles [35]. This classification is characterised by considerations related to how the use of an autonomous vehicle affects the privacy of the individual driving it.

(1) the location of the autonomous vehicle control system (ADS);
(2) the collection of data from external sources within the autonomous vehicle;
(3) the transmission of internal data gathered via the autonomous vehicle to third parties.

Each type of autonomous vehicle integrates internal data from the vehicle and external data pertaining to the environment outside the vehicle to enable autonomous driving. Internal data may encompass eye or head movement, heart rate detection, and its value, as well as the user's respiratory rate. Conversely, external data refers to the environment in which the autonomous vehicle operates, comprising data on the roadway (traffic intensity, road surface condition), meteorological conditions, and the time of day.

An autonomous vehicle is a connected vehicle [78]. The literature indicates that connected vehicles utilise networking technologies to facilitate communication among vehicles while also leveraging vehicle intelligence. According to the McKinsey Report [8], connected vehicles are increasingly prevalent in the automotive sector and are characterised by the aggregation of substantial volumes of data. The capacity of automobile manufacturers to collect data is expected to escalate as their vehicles become more connected. Projections suggest that by 2030, 95% of newly manufactured vehicles will be connected.

Interdependent vehicles utilise a dedicated communication network specifically designed for such vehicles, through which they transmit data originating from the vehicle and receive data from external sources. The Internet-of-Vehicles network ('IoV' network) refers to network technologies capable of supporting communication channels that facilitate interactions between vehicles [106]. An open and integrated IoV network system, characterised by high management, control, and operability capabilities, encompasses a multitude of users [17]. Examples of potential services and applications within the IoV network include real-time security measures and analysis of the operational status of vehicles, passenger entertainment solutions, and secure communication tailored for vehicles [15]. The vehicle itself constitutes an element of a mobile electronic communications network as defined in Article 2(1) of Directive (EU) 2018/1972 [21], functioning in a capacity akin to that of a router.

7.2.3 Legal Nature of Data Processed in Connection with the Movement of an Autonomous Vehicle

If an autonomous vehicle is to operate on European roads, the data in question may be protected under the General Data Protection Regulation (GDPR), European Union vehicle safety regulations, the confidentiality of electronic communications, as well as specific provisions established within the national legal frameworks of the Member States (e.g., refer to the Amendment to the German Road Traffic Act [34]).

The fundamental premise for the application of the GDPR is the recognition that the data processed can identify, or enable the identification of, a natural person. As stipulated in Recital 26 of the GDPR, when assessing whether a natural person can be identified, all reasonably foreseeable means—including the retrieval of records pertinent to the same individual—that are likely to be employed by the data controller or another party to directly or indirectly ascertain the identity of the natural person must be taken into account. Nevertheless, the GDPR does not extend to information that is not related to an identified or identifiable natural person, nor to personal data that has been anonymised to the extent that the data subjects cannot, or can no longer, be identified (refer to Recital 26 of the GDPR). Conversely, when data is solely pseudonymised, it continues to be classified as personal data, as pseudonymisation is reversible (see the guidelines of the European Data Protection Board (EDPB) on pseudonymisation, 2025, recital 19, [24]). Thus, within the context of the definitions articulated in Article 4(1) of the GDPR, an identifiable individual is designated as the data subject, specifically in this instance, the operator of an automated vehicle.

The definition specified in Article 4(1) of the GDPR pertains to natural persons, including individuals engaged in independent business activities. The concept of personal data encompasses all forms of information, including messages, statements, and presentations, expressed and recorded through any medium, irrespective of the manner and extent of dissemination or acquisition, which may relate to various facets of a person, encompassing their personal and material relationships, professional life, private life, education, knowledge, or character traits. According to Recital 26 of the GDPR, the exclusions encompass factors such as cost, time, and the technology employed to identify a natural person. In instances of death, data that identifies a natural person ceases to be classified as personal data, as inferred from the interpretation of Recital 27 of the GDPR. Article 4(1) of the GDPR does not provide an exhaustive list of personal data (for instance, name and surname, address, IP address, email address). Information that does not qualify as personal data consists of individual information with a considerable degree of generality (for example, street name and house number, salary amount). Nonetheless, the aforementioned information may be regarded as personal data if it is aggregated with additional information that facilitates the identification of a particular natural person.

In the context of case law, a legal position has been established regarding the classification of IP addresses as 'personal data'. An IP address consists of a unique series of digits that facilitates the mutual recognition of individual participants in communications conducted via any resource connected to the Internet, such as computers, network cards, and modems. Each device has its own IP address. The interaction between computers within the network is governed by protocols, defined as a set of rules and standards (e.g., TCP/IP) [61]. While the identification of the computer from which a connection is initiated does not allow for the identification of the natural person operating that computer, the potential for identification arises in relation to data reflecting the actual circumstances surrounding the IP address (Judgement II SA/Wa 1598/09). In instances where the administrator can reasonably be expected to employ methods to ascertain the individual behind a particular IP address, such addresses should be classified as personal data (Judgement, C-582/14). Furthermore, if an IP address is allocated for an extended duration or permanently to a specific device, and this device is associated with a designated user, it should be recognised as belonging to the data of that individual (Judgement, I OSK 1079/10).

Recital 30 of the GDPR specifies that the use of anonymous identifiers, including strings of numbers or characters typically found within cookies to establish their uniqueness, renders cookies as personal data according to Article 4, point 1 of the GDPR. Consequently, cookies, such as advertising and targeting cookies, as well as analytics cookies, should be classified as data that enables the identification of a natural person. The term "cookies" refers to small text files that are automatically created and transmitted by websites upon visitation. The information stored in these files remains on the computer or other electronic devices (such as tablets and smartphones) and can only be accessed by the originating website.

The European Union legislator has delineated three distinct categories of personal data. The first category, as defined in Article 4, paragraph 1 of the GDPR, includes all personal data that does not fall under the provisions of Article 9 and Article 10 of the GDPR. The second category comprises special categories of personal data, which are enumerated in a closed list found in Article 9, paragraph 1 of the GDPR. The third category consists of data pertaining to criminal records, as specified in Article 10, paragraph 1 of the GDPR. Furthermore, regarding the criteria concerning personal data, the EU legislator identifies the scope of personal data, to be interpreted as a collection of categories of personal data that are processed in a particular case. Based on the criterion concerning the scope of personal data, it is possible to determine the volume of personal data being processed.

An autonomous vehicle, equipped with various sensors and cameras, including, among others, a heart rate sensor and an eye-tracking camera, processes data. The information collected from these sensors and cameras, which are installed both outside and within the vehicle, facilitates the identification of the natural person using the vehicle. Consequently, this data constitutes personal data as defined in Article 4, point 1 of the GDPR. Whether this personal data also qualifies as health data, pursuant to Article 9, paragraph 1 of the GDPR, depends on the specific circumstances. For instance, if, while in a moving vehicle, the sensors monitoring heart rate and eye movements detect a decrease in previously established measurement values, such data will be transmitted to the emergency notification point regarding the potential health risk and implications for road safety. In this scenario, the scope of the data transmitted pertains to the health of the individual within the vehicle, whether the driver or the passenger. Data regarding the user's heart rate is consistently considered health data among all transmitted data. Therefore, individual sensor data can be used to infer the user's health status.

7.2.4 Legal Aspects of Processing Personal Data in a Value-Creating Ecosystem for Autonomous Vehicles, Following Established Digital Mobility Models

First and foremost, it is essential to determine whether the subject of processing pertains to information related to an 'identified person' or an 'identifiable person' as defined in Article 4, point 1 of the GDPR. Only in relation to this scope of personal data should the criteria for identifying the data controller be applied, considering the purpose and means of processing that data. According to the definition provided in Article 4(7) of the GDPR, the data controller is characterised as follows: (1) a natural or legal person, public authority, agency, or other entity; (2) who, either independently or in collaboration with others, determines; (3) the purposes and means of processing personal data. In instances where the purposes and means of such processing are governed by Union law or the legislation of a Member State, the data controller may also be designated by Union law or the legislation of a Member State, or specific criteria for its designation may be established (EDPB Guidelines 2020, point 2.1., [25].

The concept of 'processing' possesses normative significance and is delineated in Article 4(2) of the GDPR. It signifies an operation or series of operations conducted on personal data or sets of personal data, whether in an automated or non-automated manner. This includes activities such as collection, recording, organisation, structuring, storage, adaptation or modification, retrieval, consultation, utilisation, disclosure through transmission, dissemination, or any other method of making data available, matching or combining, limitation, deletion, or destruction. Furthermore, the purpose of personal data processing is instrumental in ascertaining the legal foundation for such processing. Provided that the purpose of personal data processing remains unchanged and the controller (A joint controller as defined in Article 26(1) of the GDPR may also undertake this role) is appointed for this objective, other legally distinct entities engaged in processing personal data for this purpose are classified as processors, as per Article 4(8) of the GDPR. Should there be any alteration in the purpose of personal data processing, it is imperative to repeat the process of identifying the entity responsible for determining the purposes of personal data processing.

The Report on the Ethics of Connected and Automated Vehicles [30] was commissioned by the European Commission to provide support to stakeholders, including policymakers, manufacturers, and deployers of autonomous vehicles, concerning the ethical development and regulation of such vehicles in the future. The document outlines twenty recommendations based on the fundamental ethical and legal principles set forth in the EU Treaties and the Charter of Fundamental Rights of the EU [13]. These principles are crafted to ensure that the deployment of autonomous and automated vehicles achieves the desired outcomes in terms of road safety, privacy, fairness, transparency, and accountability. It states that manufacturers of autonomous vehicles and their deployers should be recognised as controllers within the definition provided in Article 4(8) of the GDPR, as they determine the purposes and means of processing personal data. Depending on the specific circumstances, the operator of a fleet of autonomous vehicles may take on the role of a sole controller—for example, in the context of entering into a contract for the provision of transport services with a natural person—or that of a joint controller alongside the manufacturer of such vehicles. As a result, pursuant to Article 26(1) of the GDPR, the joint controllers must, through mutual agreement, clearly outline their obligations with respect to compliance with the GDPR [69].

Among the six principles outlined in Article 5 of the GDPR, the principles of lawfulness, fairness, and transparency; purpose limitation; and data minimisation are particularly significant regarding data collection in the present case. Identifying the purpose of processing personal data (excluding data specified in Article 9(1) and Article 10(1) of the GDPR) is crucial for several reasons. Firstly, it forms one of the criteria, in conjunction with the method of processing and the designation of roles, for identifying the data controller. Secondly, the purpose of processing aids in appropriately adjusting the scope of personal data processed, thereby implementing the principle of data minimisation as articulated in Article 5(1)(c) of the GDPR. Thirdly, the purpose of processing is essential for determining the legal basis for the processing of ordinary personal data. It is vital to note that, in relation to the use of an autonomous vehicle, the personal data of its users (whether drivers or passengers) may be collected for various purposes.

In this context, it is essential to initially highlight the primary recommendation outlined in the 2020 Report. Consequently, it is imperative to ensure that the utilisation of autonomous vehicles aids in mitigating harm to road users. The report asserts that, 'In line with the principle of non-maleficence, a minimal requirement for manufacturers and deployers is to ensure that CAVs decrease, or at least do not increase, the amount of physical harm incurred by users of CAVs or other road users that are in interaction with CAVs, compared to the harm that is inflicted on these groups by an appropriately calculated benchmark based on conventional driving'.

In the context of personal data processing, the 2020 Report indicates how to ensure road safety through 'continuous monitoring of the safety of autonomous and connected vehicles'. Data collection is a significant element in enhancing safety, which serves as the basis for limiting privacy in the interest of protecting the public, particularly road users. Without the collection of data, conducting comparisons or performing analyses is not feasible. Article 1(1)(g) of the German Autonomous Driving Act (2021) enumerates 13 types of data that the owner of an autonomous vehicle is mandated to retain and, upon request, provide to the authorities in accordance with privacy regulations. The obligation to retain the aforementioned data arises in the event of specific occurrences, which may include (a) interventions by the technical manager, (b) scenarios of events on the road, specifically accidents and near-accident scenarios, unplanned lane changes or turns, and malfunctions while the vehicle is in operation [91].

As users of autonomous vehicles become progressively less engaged in the intricate task of driving, their attention span may diminish. The implementation of driver monitoring systems is essential to ascertain whether the user remains focused on the task of driving at Levels 2 and 3 of driving automation. Collecting data regarding the attention span of the vehicle user could prove beneficial in ensuring road safety. For instance, systems that monitor the user's ocular or cranial movements, heart rate, or respiratory rate may assist in evaluating the user's concentration on driving.

The data collected in this manner not only pertains to the user's concentration but also provides insights into their health. Vehicles classified as automation levels 4 and 5 may be outfitted with monitoring systems, facilitating the collection of more personal health-related information. Such data may be of interest to various stakeholders, including insurers, medical practitioners, and advertisers. According to the McKinsey Report [9], several insurers have adjusted insurance rates based on driving behaviours, while certain municipalities utilise sensory data to identify road surface deficiencies. Furthermore, numerous media agencies have broadened their advertising reach through innovative touchpoints both within and beyond vehicles.

In summary, an automated vehicle has the potential to collect substantial amounts of health data from its users. The 2020 Report encompasses Recommendation 8, which aims to empower users, investigate informed consent options, and establish related industry best practices. Article 7 of the GDPR explicitly prohibits the coercion of consent. Manufacturers of autonomous and connected vehicles, as well as entities implementing solutions for these vehicles—particularly mobility service providers—are required to comply with this provision and offer adaptable tools for managing consent [39]. In this regard, public authorities are tasked with overseeing the implementation and enforcement of this requirement.

The findings derived from surveys conducted with drivers of vehicles integrated into the IoV network concerning the experiences associated with the collection and utilisation of personal data by vehicle manufacturers and collaborating entities [23] indicate, among other considerations, that the volume, context, and lifecycle of data, as well as the insufficient information on this matter, profoundly restrict the capacity of drivers to actualise their rights pertaining to this data. These instances illustrate situations in which drivers, lacking awareness of the data being collected by their vehicles, are unable to furnish informed consent.

In instances involving data related to the user's geographic location, the use of the vehicle by the user, and the journey's duration, it is essential to consider the grounds for legitimising processing as articulated in Article 6(1)(a), (b), and (c) of the GDPR. For instance, an operator managing a fleet of autonomous vehicles must ascertain who has rented the vehicle, the duration of usage, the distance covered, the location where the vehicle was parked upon completion of the journey, and the renter's details in order to impose charges for the vehicle's use. Consequently, processing this data is necessary for fulfilling contractual obligations towards users of these vehicles. Nevertheless, should the fleet operator intend to collect additional data that is not essential for the establishment and execution of the rental contract for such a vehicle, it is imperative to secure prior consent for processing such information.

Considering Article 6(3) of Regulation (EU) 2019/2144 [81], data derived from driver drowsiness and attention warning systems may be collected if deemed 'necessary concerning the purposes for which it was collected'. The GDPR delineates the limitations surrounding the collection of health data. Specifically, Article 9(1) of the GDPR asserts that the acquisition of health data is generally prohibited unless the exceptions outlined in Article 9(2) of the GDPR are applicable. Notably, the exception detailed in Article 9(2)(h) of the GDPR permits the processing of data when it is, inter alia, necessary for medical diagnosis. This exception should be interpreted in conjunction with Article 9(3) of the GDPR and Recital 53 of the GDPR, which stipulates that such data must be processed by or under the purview of a professional bound by an obligation of professional secrecy. Given that the fleet operator is not bound by such an obligation, this exception is inapplicable. An alternative exception emerges from Article 9(2)(g) of the GDPR, which pertains to the necessity of processing data for reasons of significant public interest. This must be substantiated by Union or Member State law, which ought to be proportionate to the intended objective and uphold the essence of the right to data protection. It must also incorporate appropriate and specific measures to safeguard the fundamental rights and interests of the data subject. One could argue that road safety constitutes a matter of significant public interest. In Member States possessing provisions as referenced in Article 9(2)(g) of the GDPR, this exemption may therefore be applicable. However, should a Member State lack provisions as indicated in Article 9(2)(g) of the GDPR, the exemption delineated in Article 9(2)(a) of the GDPR may apply, contingent upon the data subject (the vehicle operator) providing consent. Furthermore, the processing must adhere to the overarching principles of processing as articulated in Article 5 of the GDPR.

Approval is granted solely upon the fulfilment of numerous technical requirements derived from various legal statutes. Regulation (EU) 2018/858 [80], in recital 62, delineates that manufacturers are obligated to adopt all necessary measures to ensure compliance with the provisions regarding the processing and transfer of personal data generated during vehicle usage. Correspondingly, recital 14 of Regulation (EU) 2019/

2144 [81] asserts that the processing of personal data, including data regarding the driver collected by event data recorders or information concerning driver drowsiness and attention or distraction, must be conducted in accordance with EU data protection legislation. Event data recorders are required to function within a closed-loop system, whereby recorded data is overwritten, preventing the identification of the vehicle or the user. Furthermore, driver drowsiness and attention warning systems, as well as advanced driver distraction warning systems, must not continuously record or retain data beyond what is essential for the purposes for which the data was collected or subsequently processed in a closed-loop system.

7.2.5 Implement a Risk-Based Approach to Safeguard the Rights of Individuals Whose Data Is Processed in Digital Servitisation Business Models, with Particular Emphasis on Autonomous Vehicles

The collection and dissemination of personal data related to the operation of autonomous vehicles necessitates strict adherence to the stipulations outlined in the General Data Protection Regulation. This requirement is particularly crucial when it involves sensitive information, such as health data. The responsibilities of the data controller, as defined in Article 4(7) of the GDPR, encompass the following categories:

(1) Establishing a legal basis for the processing of personal data relevant to each category and the objectives of such processing; in the absence of a legal basis, it is necessary to either delete or anonymise the data;
(2) Safeguarding personal data, which includes implementing security measures and permitting access to data solely in accordance with an existing legal framework;
(3) Fulfilling obligations to the individual whose data is under processing, in alignment with the rights conferred by the GDPR;
(4) Meeting the obligations owed to the national data protection authority as mandated by law.

Before starting the processing activities, the data controller should adopt a systematic, multi-step approach. First, it is essential to carry out an assessment of the potential impact of the planned processing operations on the protection of personal data [69].

The European Union legislator, distinguishing between two levels of risk regarding the infringement of the rights and freedoms of natural persons in the provisions of the General Data Protection Regulation (GDPR)—namely, ordinary risk and high risk—specifies in Article 35 the conditions under which the risk may be classified as high. Such risks can arise from the inherent characteristics of artificial intelligence operations, particularly in contexts such as surveillance and profiling

[19]. In accordance with Article 35(3) in conjunction with paragraph 1 of the GDPR, the European Union legislator provides an indicative list of occurrences that may lead to a high risk of infringing upon the rights and freedoms of natural persons.

Among the identified occurrences that may pose a high risk of infringing upon the rights and freedoms of natural persons, thereby necessitating an examination of the intended processing of personal data, the European Union legislator outlines: (1) A systematic and comprehensive evaluation of personal factors related to natural persons, relying on automated processing, including profiling, which serves as the basis for decisions that have legal effects for a natural person or significantly impact a natural person in a comparable manner; (2) Extensive processing of special categories of personal data as outlined in Article 9(1) of the GDPR, or personal data concerning criminal convictions and offences, as referenced in Article 10 of the GDPR; or (3) Systematic and large-scale surveillance of publicly accessible areas.

The European Union legislator, as outlined in Article 35(4) of the GDPR, has empowered the national data protection authority, which includes the Polish President of the Personal Data Protection Office ('PDPO'), to develop and disseminate a compilation of types of processing operations that are subject to the obligation of conducting a data protection impact assessment ('DPIA'). Consequently, the Polish President of the PDPO has the authority to independently identify events that pose a significant risk to the rights or freedoms of individuals. The data protection authority is mandated to establish and publish a comprehensive list of types of processing operations governed by the requirement to conduct a data protection impact assessment (refer to Article 35(4) of the GDPR). Additionally, it is required to forward these lists to the European Data Protection Board ('EDPB'), as referenced in Article 68 of the GDPR. Furthermore, the data protection authority may formulate and disclose a list of types of processing operations exempt from the obligation to conduct a data protection impact assessment. In cases where the lists specified in Article 35(4) and (5) of the GDPR encompass processing activities related to the provision of goods or services to data subjects, or to the surveillance of their behaviour across multiple Member States, or that could substantially impact the unrestricted flow of personal data within the Union, the relevant supervisory authority shall implement the consistency mechanism delineated in Article 63 of the GDPR prior to adopting such lists. For this purpose, the Polish President of the PDPO published a notification on 17 June 2019 regarding the list of types of personal data processing operations necessitating an assessment of the impact of such processing on their protection (2019). The types of personal data processing operations delineated in this list necessitate an evaluation of the effects of processing on their protection, as referenced in Article 35(4) of the GDPR. However, there may be instances where a data controller determines that processing meets only one of these criteria, thereby necessitating a DPIA. The greater the number of criteria that the processing fulfils, the more probable it is that there will be a significant risk to the rights and freedoms of data subjects; consequently, irrespective of the measures the controller intends to implement, a DPIA will be obligatory.

The concept of personal data protection by design must be established at the initial stage of identifying the entity responsible for processing personal data. This pertains specifically to assessing whether a particular entity fulfils the criteria of a controller, as defined in Article 4, point 7 of the GDPR. In order to mitigate privacy infringements, it is essential that privacy safeguards are integrated, to the greatest extent possible, during the design phase of autonomous vehicles. The assurance of privacy significantly enhances trust in autonomous vehicles [94].

The technical and organisational measures specified by the EU legislator in Article 25(1) of the GDPR are designed to facilitate the implementation of the personal data protection principles outlined in Article 5(1) of the GDPR. In this provision, the EU legislator employs the phrase 'such as data minimisation'. The principle of data minimisation is articulated in Article 5(1)(c) of the GDPR. According to this principle, personal data must be adequate, relevant, and limited to what is strictly necessary for the purposes for which they are processed. Furthermore, the provision of Article 25(1) of the GDPR should be interpreted more expansively than merely the implementation of the principle of data minimisation; it encompasses the realisation of additional principles of personal data protection. The principle of data minimisation is interconnected with the principles of lawfulness, fairness, and transparency, as expressed in Article 5(1) of the GDPR. Moreover, it follows from the content of Article 25 (1) of the GDPR that the range of measures instituted to implement the principle of personal data protection at the design stage is aimed at providing 'the processing with the necessary safeguards to meet the requirements of this Regulation and protect the rights of data subjects' (as stated in Article 25 paragraph 1 of the GDPR in fine).

In accordance with Article 25(3) of the GDPR, the requirement to implement technical or organisational measures to safeguard personal data from the outset may be fulfilled through the introduction of an officially recognised certification mechanism, as detailed in Article 42 of the GDPR.

The obligations to provide technical and organisational measures, as articulated in Article 24(1) of the GDPR and Article 25(1) of the GDPR, warrant comprehensive analysis in conjunction with Article 32, paragraph 1, of the GDPR. The subjective scope encompassed in the norm articulated in Article 32(1) of the GDPR includes both the data controller and the data processor. This provision imposes a duty to implement technical and organisational measures that correspond to the risk of infringing upon the rights or freedoms of natural persons. An interpretation of Article 32, paragraph 1, of the GDPR leads to the conclusion that the criteria distinguishing the legal obligation to ensure the security of personal data—namely, the state of technical knowledge, the cost of implementation, and the nature, scope, context, and purposes of processing—constitute essential components for analysing the risk of violation of the rights and freedoms of natural persons. In this provision, the EU legislator specifies the duty to implement technical and organisational measures tailored to the risk of infringing upon the rights or freedoms of natural persons. Based on the aforementioned provision, both the data controller and the data processor are required to implement such measures while considering the state of technical knowledge, the cost of implementation, the nature, scope, context, and purposes of processing, as well as the likelihood and severity of the risk of infringing upon the rights or freedoms of natural persons. Furthermore, both the data controller and the data processor must ensure, in accordance with paragraph 1, the security of processing to a degree that corresponds to this risk of infringing upon the rights or freedoms of natural persons. Therefore, in determining the degree of risk of violating the rights or freedoms of natural persons, the data controller should take into account the state of technical knowledge, the cost of implementation, and the nature, scope, context, and purposes of processing.

7.3 Legal Challenges Associated with Digital Platforms in the Processing of Customer Data

7.3.1 Dominating Role of Digital Platforms in the Processing of Personal Data

The legal challenges associated with processing personal data through digital platforms are global in nature. For instance, Dignam [20] noted that in the United States, individual states such as California have enacted privacy laws designed to safeguard individuals from the misuse of their data by technology companies [2]. Additionally, the United States Congress has commenced oversight hearings concerning data misuse, while the United States Department of Justice is conducting investigations into major technology companies regarding a broad array of issues, ranging from data privacy to anti-competitive behaviour [70].

The legal challenge consists of establishing a suitable legal framework for the processing of personal data pertaining to users of digital platforms employed for the ordering and utilisation of digital mobility services. Within the European Union, the General Data Protection Regulation (GDPR) is considered an ineffective regulation, primarily due to the characteristics of digital platforms that facilitate their ability to circumvent the obligations mandated by the GDPR, thereby complicating the enforcement of the law by national regulatory authorities.

The rationale for considering the General Data Protection Regulation as not entirely effective in protecting the rights and freedoms of natural persons on digital platforms arises from several factors. These include the extensive collection of personal data from users by digital platforms, the intricate methods of service provision, the resulting complexity of data flow, and the significant impact on: (1) natural persons whose personal data is processed, (2) other entities operating within the digital sphere, (3) the execution of obligations outlined in the GDPR, as well as the limited capabilities for verifying and enforcing the stipulated obligations inherent in these provisions.

The aforementioned factors contribute to acquiring a competitive edge over individuals whose personal data is processed, collaborators, and market competitors functioning within the digital realm, in addition to national regulatory authorities that oversee the compliance with obligations stipulated in the GDPR and enforce provisions against non-compliance and inadequate execution.

Hardaker [40] articulates that digital platforms signify digital infrastructures which 'curate the interactions of interdependent complementors and users' [37]. Over the past decade, numerous digital platforms have emerged as significant entities engaging in urban governance [31], thereby influencing, modifying, and regulating essential interactions [7]. As a result, digital platforms extract value [58] and exert power [11].

To delineate the advantaged status of digital platforms in relation to the processing of personal data, the scholarly literature introduces three fundamental concepts: 'data power', 'gatekeeping power', and 'infrastructural power'. These concepts will be elucidated below.

According to Lynskey [64], 'data power' represents a multi-faceted form of authority derived from a digital platform's capacity to regulate data flows. Digital platforms possess the capability to manage such data flows as exemplifications of a two- or multi-sided market. This indicates that they operate as intermediaries between one side of the market, such as individual internet users, and others with whom they engage, including other individuals, advertisers, and service or content providers. As online platforms function as interfaces for these online interactions, they occupy a unique position to regulate the flow of information among participants in the digital ecosystem and to aggregate data concerning the actions of each of these parties within the digital realm.

In the analysis presented by Karjalainen [57], the concept of data power, which arises from the control of personal data, reduces the effectiveness of monetary sanctions established under the General Data Protection Regulation. This phenomenon is attributed to the monopolistic position that digital platforms have established through data processing. Consequently, this power undermines accountability and democratic oversight regarding data processing practices, thereby limiting individuals' control over their personal information. Furthermore, as articulated by Kerber [59], the preeminent status of digital platforms, which have developed into data monopolies, has led to a decline in privacy and hindered competition.

The concept of data power emphasises control over data as a source of power. A key factor in data power is the omnipresence of platforms in the digital environment, granting them access to large volumes of data from various applications [64]. Viewing these platforms as bearers of data power underscores that the processing of data is not inherently problematic; rather, the power that emerges from the volume and variety of data can be [64]. The concept and existence of data power also reveal the power disparity among companies [64].

van Dijck et al. [102] elucidate the intricacies associated with ownership relations, governance structures, and technical connections that remain obscure to both regulatory bodies and service users engaging with digital platforms. For instance, Facebook maintained a partnership with Uber that enabled users to request taxis through the Messenger application. In this context, it is also pertinent to note that Google is a shareholder of Uber; however, Google Maps discontinued its integration with the taxi application to ensure that its own product could attract customers from Uber's competitors. According to van Dijck et al. [102], these instances illustrate that platform companies may function as competitors in one domain while simultaneously being collaborators in another, as they can direct users toward their own branded services via rival platforms. For example, Facebook's Messenger and Instagram applications rank among the most downloaded apps in the Apple iOS App Store and are utilised as frequently as Apple's native applications that offer comparable functionalities. Through the intricate dynamics of relational platform services, technology companies are capable of managing a complex and opaque ecosystem wherein connections remain concealed from the public and largely beyond public oversight. The lack of transparency concerning centralised control over data flows engenders uncertainty for service recipients in ascertaining which entity is responsible for the processing of personal data.

Owing to the rapid growth of these enterprises during the early 2000s, legislatures have not succeeded in addressing all the challenges associated with them. Various methods for regulating data power have been extensively explored in legal scholarship; however, to this day, there remains a lack of consensus. Within the realm of competition law, the prevailing market position of these data monopolies has been analysed with respect to market power, unfair competition, and merger control [104]. In the domain of data protection law, the implementation of the GDPR in 2016 was intended to target these companies by enhancing data subject rights, increasing accountability, and imposing substantial fines. Nevertheless, the GDPR has not entirely fulfilled this objective.

According to Lynskey [64], the concept of 'gatekeeping power' should be understood in such a way that the positioning of these companies within the online market establishes them as gatekeepers of information flows among various users of the platform, including consumers, content providers, and advertisers. The gatekeepers possess the capability to collect information regarding their users, thereby gaining a competitive advantage through the volume and diversity of data they subsequently utilise for advertising purposes. According to van Dijck et al. [102], platforms therefore maintain a unique leverage regarding all types of economic transactions assigned to them as intermediaries between users, complementors, and businesses. Consequently, the gatekeeping power is evident in the management of access not only to individual services but also to the entire ecosystem. Karjalainen [57] emphasises that gatekeeping power underscores the extent of platform data processing, the capacity to amalgamate data from disparate sources, and the application of this data to regulate access to products and services. These platforms are frequently referred to as data monopolies and typically include at least Google, Facebook, Microsoft, Apple, and Amazon. Platforms may extend complimentary services to consumers, however, they primarily generate revenue by capitalising on users' personal data, which they collect and then sell to other companies for targeted advertising [60].

The concept of 'infrastructural power' regarding the behaviour of digital platforms should be understood, as indicated by Valdez [100], in relation to the routine operations of platform firms—specifically, the provision of their services—and arises from the unique position these firms occupy within the economic landscape, consequently influencing regulatory outcomes. She contends that the impact of platform companies does not merely stem from the economic growth or employment opportunities they provide (attributes of structural power) nor from their direct lobbying activities (characteristic of instrumental power). Instead, the infrastructural power of these firms is also derived from 'the specific position these firms occupy in the economy'. By 'mediat[ing] between producers and consumers of goods, services, and information (…), they create an ecosystem that depends on them' [100]. Similar to other conceptualisations of power, infrastructural power arises from the platforms' capacity to amalgamate data from various sources due to the extensive range of services each offers and the opacity surrounding these connections. The notion of infrastructural power recognises facets of both data power and gatekeeping power. This perspective underscores the collaboration among these companies. Platform power is not an attribute of a single entity but a characteristic of a comprehensive ecosystem.

7.3.2 The Manifestation of Power Through Digital Platforms

The influence of digital platforms manifests in various forms. Hardaker [40] identifies three distinct types of this influence, namely:

(1) Hard power—which involves economic and political influence, enabling corporations to shape or obstruct regulation through lobbying, advertising, and non-compliance, exemplified by Amazon's strategies regarding state tax avoidance;
(2) Soft power—which refers to the cultural influence acquired through the prestige associated with technological leadership, wherein platform CEOs shape public norms and garner support despite facing criticism;
(3) Asymmetric power, which arises from the scale of platforms, thereby allowing them to establish standards, automate operations, and control information disparities across different sectors.

Consequently, Lynskey [64] characterises 'digital power' as a form of 'soft policy power' that enables platforms to engage in and influence public and policy dialogues, as well as 'media power' through which they can regulate the content accessible to users, thereby impacting public opinion. Epstein and Robertson [26] additionally highlight the 'power and robustness' effect associated with the search engine manipulation effect.

The authors of this concept show that: (1) biased search rankings can shift the voting preferences of undecided voters by 20% or more, (2) the shift can be much higher in some demographic groups, and (3) rankings can be masked so that people show no awareness of the manipulation.

Data processing is a defining feature of the platforms and a primary factor in their power [102]. Data protection laws target this cornerstone of the platform economy. Understanding the various power dynamics that characterise the digital platform environment provides valuable tools to identify legal problems and solutions, as well as to analyse the effectiveness of different regulatory interventions in the field.

Data power may have a detrimental effect on competition between companies and harm individuals subject to this power [64].

7.3.3 Inadequate Legal Remedies to Ensure Adherence to Obligations Regarding the Processing of Personal Data

A significant endeavour to address data power was the implementation of the EU General Data Protection Regulation in 2016. While the GDPR is applicable to all forms of data processing regardless of scale, a key motivation underlying the Regulation was to limit the extensive online personal data processing undertaken by platforms [57].

The capability to process substantial volumes of personal data constitutes a pivotal factor in establishing data power among digital platforms. The perspective articulated by Karjalainen [57] warrants consideration, asserting that data power does not fully align with the European Union's data protection legal framework. It is imperative to underscore, primarily, the deficiencies in compliance regarding the principles of personal data processing, which encompass: the principle of lawfulness, fairness, and transparency (Article 5(1)(a) of the GDPR),the principle of purpose limitation (Article 5(1)(b) of the GDPR); the principle of data minimisation (Article 5(1)(c) of the GDPR); and the principle of accountability, as delineated in Article 5(2) of the GDPR. Furthermore, an additional concern lies in the bases for processing personal data, including the processing of such data predicated on consent, alongside the exercise of the rights of natural persons whose personal data are processed [67].

The impact of these dynamics on competitive markets and individual privacy constitutes a substantial impetus to pursue effective remedies against data power companies. Nevertheless, the legal framework and its enforcement efforts have not consistently demonstrated effectiveness. As noted by Karjalainen [57], these difficulties stem from an inadequate understanding of the influence exerted by these platforms.

Consequently, the principal challenge in regulating data power is the identification and remediation of the adverse effects linked to the influence wielded by digital platforms, which stems from their access to personal data. The acquisition of personal data and its ensuing processing for their objectives carries multifaceted implications for digital platforms.

First and foremost, it is imperative to delineate the scope of personal data processing, followed by subsequent analysis. The services provided by digital platforms are predicated upon the personalisation and profiling of users in relation to the services extended to them and their engagement on the Internet. Numerous services available through online platforms are offered at no cost; however, they are financed by targeted advertising that relies on highly detailed user profiles developed from behavioural data [88]. Furthermore, as noted by Karjalainen [57], a characteristic of digital platforms is their ability to offer multiple services and to amalgamate data among them. For instance, the entity recently rebranded as Meta, formerly known as Facebook, operates Facebook, Messenger, Instagram, and WhatsApp. This extensive aggregation of personal data provides companies engaged in data processing with a substantial reservoir of information and insights. These companies likely possess access to more data than any other organisation in contemporary society and have the capacity to analyse this data to derive information about the data subjects. Such access to vast quantities of data represents one of the defining features of data power and constitutes one of the most significant factors contributing to its generation [57].

Online platforms have access to knowledge and information that is unavailable to other entities. This gives them control over the production of knowledge and the external disclosure of that knowledge. In a scenario where no entities other than platforms possess comparable amounts of data and processing power, details about the nature of the data, and, more importantly, the conclusions drawn from it, remain undisclosed to the public. Because only a limited number of operators have access to this information, each holding only portions of it, the rest of society becomes reliant on them for both the provision of free and valuable services and the processing of that information for alternative purposes [107]. The concentration of this power within a select few private companies signifies a fundamental and extraordinary transformation in the traditional division of learning within an information society [107]. As articulated by Pasquale [75], the automated decision-making processes regarding the implications derived from the data and the access granted to it render the processing opaque and subject to a lesser degree of democratic control and public oversight compared to human authority over information resources.

Karjalainen [57] argues that the monopoly over data processing generates both economic and financial power for platform companies. The direct correlation between data processing and profit establishes a significant incentive for these platforms to persist in data processing, posing a challenge for the enforcement of data protection if the profitable processing fails to adhere to legal standards [27]. Monetary sanctions, in isolation, may not sufficiently motivate compliance if the penalties do not reflect the potential financial advantages of non-compliance.

The General Data Protection Regulation (GDPR) stipulates maximum penalties of 20 million Euros or up to 4% of a company's global annual revenue (GDPR Article 83(6)). In the initial three years following the enactment of the GDPR, platforms have already undergone supervisory inspections, and the first penalties have been imposed. The most significant fines have been levied against Amazon in Luxembourg (746 million Euros), WhatsApp in Ireland (225 million Euros), and Google in France (50 million Euros). Nevertheless, many of these penalties are currently subject to appeal and have not reached finality [57]. Although the GDPR allows for alternative measures, such as the prohibition of data processing (Article 58(2)), these measures have yet to be implemented on a broad scale against the platforms. An ongoing concern regarding GDPR enforcement has been the challenge of reconciling penalties with the potential financial benefits derived from data processing. The GDPR may not sufficiently compel data operators to adhere to its mandates if compliance necessitates the discontinuation or reduction of targeted advertising, which currently funds these services [27]. According to Karjalainen [57], fines should be substantial enough to offset the considerable profits generated, and effective enforcement is essential to ensure that adherence becomes economically viable. Considering the economic clout of the platforms and the significant profits they accrue from data processing, fines may not serve as an effective deterrent, particularly if their amounts remain relatively low.

Another significant challenge faced by national data protection authorities is the necessity of possessing adequate capacity, which encompasses knowledge, experience, as well as technical and organisational resources, to effectively oversee the manner in which online platforms handle personal data. The operations of online platforms are intricate and take place across multiple jurisdictions, which complicates the procurement of information pertaining to personal data processing activities and the subsequent analysis of such information [36]. This situation results in a resource disparity between the entity managing the digital platform and the national data protection authority.

The solution implemented by the European Union legislator in the General Data Protection Regulation (GDPR), which encompasses the obligation of accountability and the specification of obligations predicated on the risk of infringing upon the rights and freedoms of natural persons, is not entirely effective for entities operating digital platforms. According to Van Alsenoy [101], this approach places the responsibility for understanding the details of data processing and executing all necessary risk assessments upon the companies that process data, which should be held accountable for their actions. Conversely, Van Alsenoy [101] critiques the solution based on the specification of legal obligations outlined in the GDPR, emphasising that this risk-based criterion assigns excessive responsibility and authority to entities operating platforms regarding their compliance [101]. Furthermore, in relation to digital platforms, an approach emphasising the 'sufficiency' of protection for individuals, as opposed to the 'adequacy' of risk assessment, should be adopted [77]. According to Karjalainen [57], the extensive scope of data processing undertaken by digital platforms and the substantial data power they possess complicates supervision. At the same time, the approach towards accountability and self-regulation may present challenges due to the considerable financial stakes involved in data processing.

Simultaneously, the absence of transparency in the practices related to data power processing exacerbates the complexity of monitoring data power, as well as impacts the relationship between data power and individuals.

7.3.4 Impact on Individual Behaviour

One of the key instruments employed by the General Data Protection Regulation to protect personal data is the increased control individuals have over their own data. The notion of individual control represents a well-established principle in data protection, based on the free and informed consent of the data subject to share their data for processing.

According to Karjalainen [57], the approach outlined in the General Data Protection Regulation, which is based on the control exercised by individuals over the processing of their personal data, is ineffective against entities operating digital platforms. This inefficiency can be attributed to the following reasons:

(1) The substantial volume of online data processing carried out by digital platforms significantly reduces transparency about the flow of personal data, making it unlikely for the average individual to understand and make informed choices regarding how their data is processed by various services offered through these platforms.
(2) The significance of services provided through digital platforms in the daily lives of the average individual limits their ability to make free choices due to the fear of social exclusion.

Concerning the initial point, it is crucial to emphasise that, in accordance with Article 12(1) of the GDPR, the data controller is required to implement appropriate measures to provide any information referenced in Articles 13 and 14 of the GDPR, as well as any communication under Articles 15 to 22 and 34 of the GDPR regarding data processing to the data subject in a concise, transparent, intelligible, and readily accessible manner, utilising clear and plain language, particularly for information directed specifically towards children. In this context, Esayas [27] effectively highlights the information asymmetry that exists between entities operating digital platforms and data subjects, which is evidenced by the challenges related to comprehending informational messages concerning the processing of personal data, especially regarding the informational duties outlined in Articles 13 and 14 of the GDPR. The phenomenon of information asymmetry is further exemplified by the practices of digital platforms that limit the dissemination of information regarding personal data processing to the minimum required by the GDPR, justifying this limitation based on the protection of trade secrets [38]. The failure to furnish such information indicates that the data controller, as the entity overseeing an internet platform, has contravened the transparency obligations mandated by the GDPR. From the viewpoint of the data subject, this also implies that the individual is unable to obtain essential information regarding the processing of their personal data and, as a consequence, is hindered from making informed decisions in this respect.

Some digital services are designed to enable users to continually examine the digital traces left by those with whom they are connected via social media. For example, the intended use of Facebook is to 'connect and share with the people in your life' by browsing the news feed and user profiles [65]. Theorists have labelled this trend as lateral surveillance [5], participatory surveillance [3], social searching [62], and social surveillance [42]. In some cases, adherence to this social trend is compelled, driven by fear of social exclusion [105], both in the workplace and in activities outside of it.

There is a discernible trend in the literature whereby digital platforms seek to justify the extensive processing of personal data and the restricted accessibility of this data, with the quality of service provision deemed to outweigh concerns regarding privacy protection. In this context, Karjalainen [57] astutely observes that although individuals may be inclined to exchange a portion of their privacy for enhanced services, various behavioural factors inhibit them from fully addressing privacy concerns and advocating for improved data protection.

Opacity and a lack of transparency contribute to, and are indeed essential components of, the power associated with data. The absence of transparency can impede surveillance efforts, thereby safeguarding the source of the data and facilitating the continuation of existing surveillance practices [38]. Furthermore, another rationale for secrecy lies in the disparity between data processing and cultural perceptions of privacy [107]. Essentially, the specifics concerning data processing are often concealed due to fears regarding potential public disapproval [57].

In conclusion, the influence of data significantly impacts the regulatory approaches implemented by the General Data Protection Regulation. It diminishes the efficacy of financial sanctions due to the monopolistic power that digital platforms have established through extensive data processing. Furthermore, it reduces accountability and democratic oversight of data processing practices while hindering individual control over personal data because of issues of opacity and normalisation. Therefore, it seems reasonable to assert that the GDPR may not identify the most effective strategies to address the challenges of platform data processing.

7.3.5 Towards Industry Regulation of Digital Platforms

The European Union has implemented a strategy that emphasises digital platforms as a specialised industry, potentially in response to critiques regarding the efficacy of data protection and competition legislation in addressing platform-related issues. The Digital Services Act package, which consists of Regulation (EU) 2022/1925 ('Digital Markets Act', 'DMA') [83] and Regulation (EU) 2022/2065 ('Digital Services Act', 'DSA') [84], strengthens the regulations governing digital services and upholds fundamental rights while fostering fair competition.

The Digital Markets Act has been specifically designed to promote fair competition and mitigate anti-competitive behaviours, thereby establishing a more balanced environment within a digital ecosystem largely controlled by gatekeepers. It governs enterprises that offer at least one of the designated 'Core Platform Services' ('CPSs') to numerous end users within the European Union. Although the responsibilities outlined in the DMA primarily pertain to the specifically identified 'gatekeepers', they may also present considerable opportunities for entrepreneurs.

The Digital Markets Act was officially adopted by the European Parliament and the Council on 14 September 2022, became effective on 1 November 2022, and commenced application on 2 May 2023. Article 1(2) of the Digital Markets Act stipulates that it shall be applicable to core platform services provided or offered by gatekeepers to business users established within the Union or to end users established or located within the Union, regardless of the place of establishment or residence of the gatekeepers and regardless of the governing law applicable to the provision of the service. As defined in the DMA, a 'gatekeeper' refers to an undertaking that offers one of the ten core platform services delineated in Article 2(2), which has been designated by the Commission to assume this role in accordance with Article 3 of the regulation. In September 2023, the Commission designated six companies as Gatekeepers: Alphabet, Amazon, Apple, ByteDance, Meta, and Microsoft. The designation process for specific providers of core platform services remains ongoing [29].

Once the Commission has designated a company as a provider of core platform services, such an undertaking is subject to several restrictions and obligations, particularly those related to the processing of personal data. The gatekeeper faces the following limitations:

(1) The processing of personal data of end users accessing the services of third parties utilising the core platform services of the gatekeeper is permitted solely for the purpose of providing online advertising services.
(2) The combination of personal data from the relevant core platform service with personal data from additional core platform services, or any other services offered by the gatekeeper or third-party services, is prohibited.
(3) The cross-utilisation of personal data from the relevant core platform service in other independently provided services by the gatekeeper, including other core platform services, and vice versa, is not allowed.
(4) Registration of end users in additional services of the gatekeeper for the purpose of combining personal data is forbidden.

Such methods of processing may, however, be permitted, provided that the end user has given consent, as defined in Article 4(11) and Article 7 of the GDPR, after being presented with a specific choice.

In accordance with the obligation set out in Article 6(10) of the GDPR, the gatekeeper is required to provide business users and third parties authorised by a business user, upon their request and at no cost, with effective, high-quality, continuous, and real-time access to and use of aggregated and non-aggregated data, including personal data.

With respect to data, the Digital Markets Act stipulates provisions for interoperability (Article 7 of the DMA) and data portability, designed to prevent lock-in effects and facilitate transitions between services [63]. It is advisable to endorse Karjalainen's [57] perspective that the obligations of gatekeepers concerning personal data processing can be regarded as self-executing, as the responsibility for compliance rests with the company. This characteristic aligns the Digital Markets Act with the General Data Protection Regulation, particularly regarding its obligations concerning accountability. Furthermore, the DMA, unlike other regulations that apply to digital platforms, specifically addresses only those companies deemed to possess data power. Once a company is identified as a core platform service provider, no further assessment of potential power misuse is conducted. This methodology may help circumvent the existing challenges presented by the GDPR relating to the evaluation of actions regarding data power.

Although the General Data Protection Regulation and the Digital Markets Act address markedly distinct dimensions of data authority, it is intriguing to juxtapose their methodologies, as both laws employ nearly diametrically opposed strategies in targeting what they consider the most detrimental activities. In contrast to the GDPR, which imposes obligations on all entities irrespective of scale and bases the mitigation of these obligations on the risks associated with data processing rather than the characteristics of the company itself, the DMA initially identifies those gatekeeper entities deemed to pose risks and subsequently enacts specific obligations tailored to them [76]. The transition towards categorising digital platforms as an industrial sector, governed by its own unique regulatory framework—as demonstrated by the DMA—may represent a significant advancement in the evolution of European platform regulation. Nevertheless, given that the DMA primarily functions as a competition law instrument, it is essential to recognise that distinct data protection regulations may also be applicable to digital platforms.

The Digital Services Act emphasises the regulation of online content and the protection of users. This regulation, among other objectives, modifies and augments existing rules for intermediary service providers. Specifically, it addresses whether 'mere conduits', 'caching', or 'hosting' service providers should be held liable for the legality of the data that traverses their infrastructure. These modifications aim to tackle the challenges presented by digital platforms, thereby promoting a safer and more transparent online environment [14]. The DSA imposes obligations relating to the processing of personal data and the safeguarding of data subjects. Pursuant to Article 27(3) of the DSA, providers of online platforms are prohibited from displaying advertisements to service recipients based on profiling, as outlined in Article 4, point (4) of the GDPR, particularly concerning sensitive personal data referenced in Article 9(1) of the GDPR. Moreover, in accordance with Articles 28(2) and (3) of the DSA, providers of online platforms must refrain from displaying advertisements on their interface based on profiling, utilising the personal data of the service recipient when they are reasonably certain that the service recipient is a minor. Furthermore, adherence to the obligations outlined in this Article shall not necessitate that providers of online platforms process additional personal data to ascertain whether the service recipient is a minor.

The Digital Services Act mandates that providers of very large online platforms and very large online search engines conduct a risk assessment regarding the risks within the European Union that arise from the design or functioning of their services and associated systems, including algorithmic systems, or from the utilisation of their services. This risk assessment must be specifically tailored to their services and proportionate to the systemic risks, taking into account both their severity and likelihood. Furthermore, it should encompass, among other elements, systemic risks and any actual or foreseeable adverse effects on the exercise of fundamental rights, particularly the right to human dignity as enshrined in Article 1 of the Charter of Fundamental Rights of the EU (Charter 2007), the right to respect for private and family life as outlined in Article 7 of the Charter (2007), and the right to the protection of personal data as stipulated in Article 8 of the Charter (2007).

The legal aspects pertinent to the protection of personal data are also outlined in Regulation (EU) 2023/2854 ('Data Act') [85] and Regulation (EU) 2022/868 ('Data Governance Act') [82]. The Data Act serves to complement the Data Governance

Act, which is recognised as the first deliverable under the European Strategy for Data [16]. The Data Governance Act became effective in September 2023. While the Data Governance Act promotes confidence in voluntary data-sharing mechanisms, the Data Act provides legal clarity regarding access to and utilisation of data.

In conjunction with various policy measures and funding opportunities, these two regulations will significantly contribute to establishing a European Union single market for data, positioning Europe as a leader in the data economy by effectively utilising the vast potential of the continually increasing volumes of data, particularly industrial data, for the advancement of the European economy and society. The Data Act is set to take effect on 12 September 2025.

The Data Act represents a legislative framework aimed at enhancing the data economy within the European Union and promoting a competitive data market. This is achieved by facilitating greater accessibility and usability of data, particularly industrial data, thus encouraging data-driven innovation and improving data availability. To realise these objectives, the Data Act establishes principles of fairness regarding the distribution of data value among various participants in the data economy. Furthermore, it outlines the criteria under which different entities may utilise specific data.

In recent years, there has been a significant increase in the availability of internet-connected products, referred to as 'connected products', within the European market. These products collectively form a network known as the Internet of Things (IoT), which greatly enhances the volume of data available for reuse within the European Union (EU). This development presents considerable potential for innovation and competitiveness in the EU.

The Data Act provides users of connected products—whether businesses or individuals who own, lease, or rent such products—with enhanced control over the data they generate, all while preserving incentives for those who invest in data technologies. Furthermore, it establishes general conditions that govern the scenarios in which a business is legally compelled to share data with another business.

The Data Act incorporates measures designed to enhance fairness and competition within the European cloud market, as well as to safeguard companies from unjust contractual conditions pertaining to data sharing enforced by more dominant entities. Furthermore, it establishes a mechanism enabling public sector organisations to solicit data from businesses in cases of exceptional necessity, such as public emergencies, and delineates clear protocols for the formulation of such requests. Additionally, it implements safeguards to prohibit government bodies from third countries from accessing non-personal data in instances where such access would contravene EU or national law [28].

Numerous regulatory frameworks, including the General Data Protection Regulation (GDPR), are currently in place and impose rigorous guidelines regarding the use of personal data. At present, four legislative acts have been developed that will define the standards for Europe: the Data Act, the Data Governance Act, the Digital Services Act (DSA), and the Digital Markets Act (DMA), which are expected to take effect from September 2025.

7.4 Legal Aspects of Electronic Communication Confidentiality in Relation to the Provision of Digital Mobility Services

To obtain digital mobility services, access to the internet is essential. Consequently, all individuals utilising these services must ensure they have arranged for internet access. These services are classified as publicly available electronic communication services (refer to Article 2(4) of the Directive (EU) 2018/1972 ('Electronic Communications Code')). By entering into a contract for the provision of digital mobility services and utilising them, the customer is entitled to the protections arising from the principles of confidentiality inherent in electronic communications. In this context, it is imperative to elucidate the legal sources from which such protections are derived.

The principal normative act is Directive 2002/58/EC, commonly referred to as the 'ePrivacy Directive'. As articulated in Article 1(1) of this Directive, its intent is to safeguard fundamental rights and freedoms, particularly the right to privacy, in relation to the processing of personal data within electronic communications. Article 3 of this directive specifies that it is applicable to the processing of personal data associated with the provision of publicly available electronic communications services within public communications networks in the European Union. The European Court of Justice determined in the judgement C 207/16 dated 2 October 2018, that the ePrivacy Directive must be considered as governing the activities of the providers of such services.

The ePrivacy Directive complements the General Data Protection Regulation (GDPR) by establishing a more specific framework. Article 1(2) of the ePrivacy Directive includes a direct reference to Directive 95/46/EC, which was repealed on 25 May 2018. Nonetheless, according to Article 94(2) of the GDPR, references to the repealed Directive shall be interpreted as references to the GDPR.

In the context of the relationship between the GDPR and the ePrivacy Directive, Article 95 of the GDPR clarifies that it does not impose additional obligations on natural or legal persons regarding processing linked to the provision of publicly available electronic communications services in public communications networks (see Article 2(1) of the Electronic Communications Code), in relation to matters for which they are subject to specific obligations with the same objective outlined in the ePrivacy Directive.

The intentions of the European Union legislator regarding the scope of application of both regulations are articulated in recital 173 of the GDPR. It is evident from this recital that the GDPR is intended to apply to all matters related to the protection of fundamental rights and freedoms in connection with the processing of personal data not governed by specific obligations with similar objectives as those outlined in the ePrivacy Directive. In terms of obligations, it is essential to consider the responsibilities of the data controller and the corresponding rights of natural persons.

It can be concluded that the legislator of the European Union intended to indicate that the ePrivacy Directive constitutes a regulation specific (lex specialis) to the General Data Protection Regulation, when recital 173 of the GDPR is analysed in conjunction with Article 95 of the GDPR. The relationship between these regulations is further elucidated by referring to the ePrivacy Directive as lex specialis, as evidenced in the judgment of the European Court of Justice in Joined Cases C-511/18, C-512/18 and C-520/18 dated 6 October 2020. A similar conclusion can be drawn through an analysis of Recital 10 of the ePrivacy Directive, which necessitates interpretation in conjunction with the first sentence of Article 94(2) of the GDPR. Furthermore, Recital 10 of the ePrivacy Directive stipulates that the GDPR can only be applicable in instances where the ePrivacy Directive has not addressed matters relating to the protection of fundamental rights and freedoms.

The European Data Protection Board ('EDPB') adopted Opinion 5/2019 on 12 March 2019 [72], which addresses the interplay between the ePrivacy Directive and the General Data Protection Regulation, particularly concerning the competencies, responsibilities, and authorities of Data Protection Authorities ('DPAs'). In its analysis, it indicated, inter alia, that the material scope of the General Data Protection Regulation ('GDPR') encompasses any form of processing of personal data, regardless of the technology used. Conversely, the GDPR does not apply when the subject of the processing does not involve personal data and when the processing of personal data falls outside the material and territorial scope outlined in Articles 2(2) and (3) of the GDPR, as specified in the EDPB Guidelines 3/2018 [6].

From the perspective of the material scope of the ePrivacy Directive, it is important to note that the Directive encompasses electronic communications services offered over electronic communications networks. Moreover, electronic communications services and networks are accessible to the public within the European Union.

In the context of the application scope of the ePrivacy Directive, it is imperative to emphasise that its primary objective is to protect the fundamental rights and freedoms of individuals utilising electronic communications networks. However, the scope of electronic communications services extends further. Accordingly, Articles 5(3) and 13 of the ePrivacy Directive apply not only to providers of electronic communications services but also to website operators (for instance, regarding cookies) and other entities (for example, in relation to direct marketing).

In accordance with Opinion 1/2008 issued by the Article 29 Working Party [71] and in conjunction with Article 94(2) of the General Data Protection Regulation (GDPR), search engine services that store cookies on a user's device or access those cookies are included within the extensive material scope delineated in Article 5(3) of the ePrivacy Directive. Under this provision, Member States are obligated to ensure that the use of electronic communications networks to store information or to access information stored in the terminal equipment of a subscriber or user is permitted solely on the condition that the subscriber or user is provided with explicit and comprehensive information in accordance with the GDPR, particularly regarding the purposes of processing, as well as the right to refuse such processing by the data controller. Furthermore, this does not impede any technical storage or access solely for the purpose of executing or facilitating the transmission of a communication over an electronic communications network, or as strictly necessary to provide an information society service explicitly requested by the subscriber or user. Moreover, regarding Article 13 of the ePrivacy Directive, which addresses unsolicited emails sent by a website operator for direct marketing purposes, these activities also fall within the extended material scope as outlined in Article 13 of the ePrivacy Directive.

The ePrivacy Directive and the General Data Protection Regulation may apply concurrently to the use of cookies. This was substantiated by the Article 29 Working Party in its Opinion 2/2010. The key issue connecting both regulations is the acknowledgement that the information collected is classified as personal data pursuant to Article 4, point 1 of the GDPR. Accordingly, both the GDPR and the ePrivacy Directive apply to the processing of such data. The rationale for enforcing both regulations is articulated in the examination of Recital 30 of the GDPR, where the European Union legislator delineates various categories of online identifiers linked to a natural person. Such online identifiers can be obtained through various methods, including end devices and applications. This recital thus references Internet Protocol (IP) addresses and cookie identifiers.

7.5 Conclusions

The provision of digital mobility services necessitates the processing of substantial volumes of data that facilitate the identification of individuals, thereby constituting personal data. The legal protection of personal data is derived from the safeguarding of fundamental values, which underpin the establishment of a legal framework in this domain. Within European Union legislation, the foundation for legal protection of personal data is anchored in the principle of human dignity. Furthermore, in the national laws of Member States, additional values, alongside human dignity, are recognised as sources of data protection regulations. For instance, the case law of the Polish Constitutional Court highlights the human right to 'informational autonomy'. In the context of Japanese legislation, the foundation for legal regulations pertaining to the protection of personal data is grounded in the concept of 'respect', which possesses a similar connotation to 'dignity'. The legal protection of personal data assumes considerable importance within the framework of the utilisation of autonomous vehicles and digital platforms.

The intricate operation of autonomous vehicles and their interaction with the environment necessitates a thorough examination of the legal framework and the nature of data concerning the right to privacy. Localisation and mapping algorithms ascertain both the global and local positioning of the vehicle while concurrently constructing a map of the surroundings utilising data from sensors and other perception sources. The integration of various sensors and camera systems in autonomous vehicles results in the processing of substantial data, which informs decision-making regarding vehicle operation. Each autonomous vehicle amalgamates internal data collected from the vehicle with external environmental information to facilitate self-driving capabilities. Internal data may include metrics such as ocular or cranial movements, heart rate, and respiratory rate. In contrast, external data encompasses factors pertinent to the vehicle's driving environment, including road conditions (such as traffic and surface status), weather, and the time of day. When operating on European roads, this data may be protected under GDPR regulations, EU vehicle safety laws, electronic communications secrecy, and specific provisions enacted by Member States' national legal systems. Autonomous vehicles, equipped with sensors and cameras, collect data that can identify the individual using the vehicle, thereby categorising this information as personal data in accordance with Article 4(1) of the GDPR. The classification of such data as health data under Article 9, paragraph 1 of the GDPR is contingent upon particular circumstances. Level 4 and 5 vehicles may incorporate monitoring systems that collect additional personal health information. The obligations of a personal data controller encompass the following responsibilities: (1) establishing a lawful basis for processing each category of personal data and its associated purpose; in the absence of such a basis, the data must be deleted or anonymised; (2) safeguarding personal data, ensuring it is secure and accessible only under an existing legal justification; (3) fulfilling the obligations owed to individuals whose data is processed in accordance with their rights under the GDPR; (4) adhering to requirements set forth by the national data protection authority as stipulated by legal provisions.

Prior to the initiation of data processing activities, it is incumbent upon the data controller to comply with a multi-step methodology. This process begins with an assessment of the potential impacts that the proposed processing activities may exert on the safeguarding of personal data. The General Data Protection Regulation cannot be unequivocally regarded as wholly effective in its capacity to protect individuals' rights and freedoms within the context of digital platforms, owing to several limitations, including the considerable accumulation of personal data from users, the intricate nature of service delivery, the resultant complex data flows, and the implications for: (1) individuals whose data is processed, (2) other stakeholders within the digital sphere, and (3) compliance with GDPR obligations, in conjunction with the limited ability to verify and enforce these obligations. To elucidate the advantageous position of digital platforms in the processing of personal data, three concepts that are frequently addressed in the literature include: "data power," "gatekeeping power," and "infrastructure power," which will be elaborated upon subsequently. Data processing is a fundamental characteristic of platforms, playing a pivotal role in augmenting their influence [102]. Consequently, data protection legislation attends to this critical aspect of the platform economy. Comprehending the various power dynamics inherent in the digital platform ecosystem equips us with essential instruments to identify legal challenges and formulate solutions, as well as to assess the effectiveness of diverse regulatory measures within this sphere.

The principal challenge in regulating data authority lies in identifying and mitigating the adverse consequences arising from the power exercised by digital platforms due to their access to personal data. This access to personal data and its subsequent processing for specific purposes carries multifaceted significance for digital platforms. Furthermore, national data protection authorities face the challenge of possessing genuine capabilities, including knowledge, experience, and both technical and organisational resources, to effectively oversee the processing of personal data by online platforms. European Union regulations, exemplified by the Digital Markets Act and the Digital Services Act, aim to strengthen the rules governing digital services, safeguard fundamental rights, and promote fair competition. Consequently, it may be inferred that the legislator within the European Union intended to indicate that the ePrivacy Directive functions as a special regulation (lex specialis) in relation to the General Data Protection Regulation. In light of the imperative to uphold human dignity and protect the human rights and fundamental freedoms of every individual, whilst also considering the diversity, intensification, and globalisation of data processing and personal data flows, personal autonomy is rooted in the right of individuals to exercise control over their personal data and its processing, thereby establishing a framework for informational autonomy. Thus, the ePrivacy Directive represents a regulation specifically tailored (lex specialis) to the GDPR, as it provides a more defined framework.

References

1. Act on the Protection of Personal Information, Act No. 57 of May 30, 2003, Consolidated Text as of April 1, 2023. https://www.japaneselawtranslation.go.jp/en/laws/view/4241/en. Accessed 24 Feb 2025
2. Adler, A.: California passes strict Internet privacy law with implications for the country. NPR, 29 June 2018. https://www.npr.org/2018/06/29/624336039/california-passes-strict-internet-privacy-law-with-implications-for-the-country. Accessed 21 Feb 2025
3. Albrechtslund, A.: Online social networking as participatory surveillance. First Monday **13**(3) (2008). http://www.uic.edu/htbin/cgiwrap/bin/ojs/index.php/fm/article/view/2142/1949. Accessed 21 Feb 2025
4. Anderson, J.M., Kalra, N., Stanley, K.D., Sorensen, P., Samaras, C., Oluwatola, T.A.: Autonomous Vehicle Technology: A Guide for Policymakers. Research Report. RAND Corporation (2016)
5. Andrejevic, M.: The work of watching one another: lateral surveillance, risk, and governance. Surveill. Soc. **2**(4), 479–497 (2005)
6. EDPB,:, Guidelines 3/2018 on the Territorial Scope of the GDPR (Article 3), Version 2.1, Adopted 12 November 2019, (2019) https://www.edpb.europa.eu/sites/default/files/files/file1/edpb_guidelines_3_2018_territorial_scope_after_public_consultation_en_1.pdf; Accessed 21 Feb 2025.
7. Barns, S.: Negotiating the platform pivot: from participatory digital ecosystems to infrastructures of everyday life. Geogr. Compass **13**(9), 1–13 (2019)
8. Bertoncello, M., Martens, C., Möller, T., Schneiderbauer, T.: Unlocking the Full Life-Cycle Value from Connected-Car Data. McKinsey & Company (2021). https://www.mckinsey.com/industries/automotive-and-assembly/our-insights/unlocking-the-full-life-cycle-value-from-connected-car-data/. Accessed 27 Jan 2025
9. Bertoncello, M., Martens, C., Möller, T., Schneiderbauer, T.: Unlocking the Full Life-Cycle Value from Connected-Car Data. McKinsey & Company (2021). https://www.mckinsey.com/industries/automotive-and-assembly/our-insights/unlocking-the-full-life-cycle-value-from-connected-car-data/. Accessed 27 Feb 2025
10. Büchi, M., et al.: The chilling effects of algorithmic profiling: mapping the issues. Comput. Law Secur. Rev. **36**, 105367 (2020)
11. Busch, C., Graef, I., Hofmann, J., Gawer, A.: Uncovering blind spots in the policy debate on platform power. Final report. European Commission (2021). https://www.pure.uvt.nl/ws/portalfiles/portal/49067689/Uncovering_blindspots_in_the_policy_debate_on_platform_power.pdf. Accessed 21 Apr 2025

12. Cabinet Office of Japan: Social Principles of Human-Centric AI, Mar 2019. file:///C:/Users/akras/Documents/Andrzej_Naukowe/Andrzej_Ksi%C4%85%C5%BCki/Springer_Legal%20Challenges%20of%20a%20Transformation%20%E2%80%93%20Servitization%20and%20Autonomous%20Vehicles/Pisanie%20ksi%C4%85%C5%BCki_14.02.2025/materia%C5%82y/Transit/Japan/data%20protection%20laws%20in%20Japan.pdf. Accessed 24 Feb 2025
13. Charter of Fundamental Rights of the European Union (2000/C 364/01). https://www.europarl.europa.eu/charter/pdf/text_en.pdf. Accessed 21 Feb 2025
14. Chawla, K.: An overview of relevant European legislation for MSMEs engaged in data sharing centre of excellence for data sharing & cloud (coe-dsc.nl) (2024). https://www.resolver.tno.nl/uuid:55acffb9-1a40-44f7-aaa2-0a6d85dc8aa6. Accessed 12 Apr 2025
15. Cheng, B.-C., Chou, C.H.-T., Chang, C.-H., Lo, K.-H., Lin, P., Hsing, T.R.: Survey on networking for internet of vehicles. J. Internet Technol. **14**(7), 1007–1019 (2013)
16. Communication from the Commission to the European Parliament, the Council, the European Economic and Social Committee and the Committee of the Regions, a European Strategy for Data, COM(2020) 66 Final, 19.02.2020
17. Contreras, J., Zeadally, S., Guerrero-Ibanez, J.A.: Internet of vehicles: architecture, protocols, and security. IEEE Internet Things J. **4662**(c), 1 (2017)
18. Convention No. 108 of the Council of Europe for the Protection of Individuals with Regard to Automatic Processing of Personal Data, Drawn Up in Strasbourg on 28 January 1981 (OJ 2003, No. 3, Item 25, as Amended)
19. de Almeida, P.G.R., Dos Santos, C.D., Farias, J.S.: Artificial intelligence regulation: a framework for governance. Ethics Inf. Technol. **23**, 505–525 (2021). https://doi.org/10.1007/s10676-021-09593-z
20. Dignam, A.: Artificial intelligence, tech corporate governance and the public interest regulatory response. Camb. J. Reg. Econ. Soc. **13**, 37–54 (2020). https://doi.org/10.1093/cjres/rsaa002
21. Directive (EU) 2018/1972 of the European Parliament and of the Council of 11 December 2018 Establishing the European Electronic Communications Code (Recast) (Text with EEA Relevance), PE/52/2018/REV/1, OJ L 321, 17.12.2018, pp. 36–214
22. Donaldson, L.: Organizational design and life-cycle of products. J. Manag. Stud. **22**(1), 25–37 (1985)
23. Dowthwaite, A., Cook, D., Cox, A.L.: Privacy preferences in automotive data collection. Transp. Res. Interdiscip. Perspect. **24**, 101022 (2024). https://doi.org/10.1016/j.trip.2024.101022
24. EDPB, 2025, Guidelines 01/2025 on Pseudonymization, Adopted on 16.01.2025. https://www.edpb.europa.eu/system/files/2025-01/edpb_guidelines_202501_pseudonymization_en.pdf. Accessed 29 Jan 2025
25. EDPB, 2021, Guidelines 07/2020 on the Concepts of Controller and Processor in the GDPR, Version 2.1., Adopted on 7 July 2021, https://www.edpb.europa.eu/system/files/2023-10/EDPB_guidelines_202007_controllerprocessor_final_en.pdf, Accessed 29 Jan 2025
26. Epstein, R., Robertson, R.E.: The search engine manipulation effect (SEME) and its possible impact on the outcomes of elections. Proc. Natl. Acad. Sci. **112**, E4512 (2015). https://doi.org/10.1073/pnas.1419828112
27. Esayas, S.Y.: Competition in (data) privacy: 'zero'-price markets, market power, and the role of competition law. Int. Data Privacy Law **8**, 187 (2018)
28. EU Commission (2025). https://www.digital-strategy.ec.europa.eu/en/factpages/data-act-explained#:~:text=The%20Data%20Act%20is%20a%20law%20designed%20to,usable%2C%20encouraging%20data-driven%20innovation%20and%20increasing%20data%20availability. Accessed 23 Jan 2025
29. European Commission (2025). https://www.ec.europa.eu/commission/presscorner/detail/en/ip_23_4328. Accessed 12 Apr 2025
30. European Commission: Ethics of connected and automated vehicles: recommendations on road safety, privacy, fairness, explainability and responsibility, 17 Sept 2020.

Directorate-General for Research and Innovation. Publications Office of the European Union. Available online http://www.op.europa.eu/en/publication-detail/-/publication/89624e2c-f98c-11ea-b44f-01aa75ed71a1/language-en. Accessed 21 Feb 2025
31. Fields, D., Bissell, D., Macrorie, R.: Platform methods: studying platform urbanism outside the black box. Urban Geogr. **41**(3), 462–468 (2020)
32. Floridi, L.: On human dignity as a foundation for the right to privacy. Philos. Technol. **29**(4), 307 (2016). https://doi.org/10.1007/s13347-016-0220-8
33. Gates, G.: When cars drive themselves. New York Times, 14 Dec 2016. https://www.nytimes.com/interactive/2016/12/14/technology/how-self-driving-cars-work.html. Accessed 21 Feb 2025
34. Gesetz zur Änderung des Straßenverkehrsgesetzes und des Pflichtversicherungsgesetzes—Gesetz zum autonomen Fahren, 12 Juli 2021. https://www.bgbl.de/xaver/bgbl/start.xav#__bgbl__%2F%2F*%5B%40attr_id%3D%27bgbl121s3108.pdf%27%5D__1738182616682. Accessed 21 Jan 2025
35. Glancy, D.J.: Privacy in autonomous vehicles. Santa Clara L Rev. **52** (2012)
36. Gonçalves, M.E.: The EU data protection reform and the challenges of big data: remaining uncertainties and ways forward. Inf. Commun. Technol. Law **26**, 90, 101 (2017)
37. Grabher, G., König, J.: Disruption, embedded. A Polanyian framing of the platform economy. Sociologica **14**(1), 95–118 (2020)
38. Grochowski, M., Jabłonowska, A., Lagioia, F., Sartor, G.: Algorithmic transparency and explainability for EU consumer protection: unwrapping the regulatory premises. Crit. Anal. Law **8**, 46 (2021)
39. EDPB, 2020, Guidelines 1/2020 on processing personal data in the context of connected vehicles and mobility related applications. Version 1.0. Adopted 28 Jan 2020. https://www.edpb.europa.eu/sites/default/files/consultation/edpb_guidelines_202001_connectedvehicles.pdf. Accessed 31 Feb 2025
40. Hardaker, S.: Platformizing structural policy instruments? Fostering (infrastructural) power in the context of digital free trade zones. Geoforum **160**, 104218 (2025). https://doi.org/10.1016/j.geoforum.2025.104218
41. Jaffe, E.: The First Look at How Google's Self-Driving Car Handles City Streets. CityLab. http://www.citylab.com/tech/2014/04/first-look-how-googles-self-driving-car-handles-city-streets/8977/. Accessed 21 Jan 2025
42. Joinson, A.N.: Looking at, looking up or keeping up with people?: motives and use of Facebook. In: Proceeding of the Twenty-Sixth Annual SIGCHI Conference on Human Factors in Computing Systems, pp. 1027–1036 (2008)
43. Judgment of the Constitutional Court of 11 April 2000, K-15/98, OTK ZU Nr 3/2000, Pos. 86
44. Judgment of the Constitutional Court of 16 December 2000, SK 26/16, OTK-A 2020, No. 69
45. Judgment of the Constitutional Court of 19 February, 2002, U 3/01, OTK ZU Nr 1/A/2002/3
46. Judgment of the Constitutional Court of 12 November 2002, SK 40/01, OTK ZU, No. 6/A/2002, Item 81
47. Judgment of the Constitutional Court of 20 November, 2002, K-41/02, OTK-A 2002, No. 6, Pos. 83
48. Judgment of the Constitutional Court of 20 April 2004, K 45/02, OTK ZU, No. 4/A/2004, Item 30
49. Judgment of the Constitutional Court of 17 June 2008, K 8/04, OTK ZU, No. 5/A/2008, Item 81
50. Judgment of the Constitutional Court of 13 December 2011, K 33/08, OTK ZU, No. 10/A/2011, Item 116
51. Judgment of the Constitutional Court of 16 December 2020, SK 26/16, OTK-A 2020, No. 69
52. Judgment of the Court (Second Chamber) of 19 October 2016, Patrick Breyer v Bundesrepublik Deutschland, Case C-582/14, ECLI:EU:C:2016:779
53. Judgment of the ECJEU of 2 October 2018, Ministerio Fiscal v. Audiencia Provincial de Tarragona, C-207/16, ECLI:EU:C:2018:788

54. Judgment of the ECJEU of 6 October 2020, La Quadrature du Net and Others v Premier Ministre and Others, Joined Cases C-511/18, C-512/18 and C-520/18, ECLI:EU:C:2018:788
55. Judgment of the Supreme Administrative Court of 9 May 2011, I OSK 1079/10, CBOSA. Accessed 21 Feb 2025
56. Judgment of the Voivodeship Administrative Court of 3 February 2010, II SA/Wa 1598/09, CBOSA. Accessed 21 Feb 2025
57. Karjalainen, T.: The battle of power: enforcing data protection law against companies holding data power. Comput. Law Secur. Rev. **47**, 105742 (2022)
58. Kenney, M., Zysman, J.: Work and value creation in the platform economy. In: Vallas, S.P., Kovalainen, A. (eds.) Work and Labor in the Digital Age (Research in the Sociology of Work, Vol. 33), pp. 13–41. Emerald Publishing Limited, Leeds (2019). https://doi.org/10.1108/S0277-283320190000033003
59. Kerber, W.: Digital markets, data, and privacy: competition law, consumer law and data protection. J. Intell. Prop. Law Pract. **11**, 856, 859 (2016)
60. Kira, B., Sinha, V., Srinivasan, S.: Regulating digital ecosystems: bridging the gap between competition policy and data protection. Ind. Corp. Change **30**, 1337, 1340 (2021)
61. Krasuski, A.: Leksykon prawa nowych technologii pod. In: Krasuski, A. (ed.) Warszawa (2021)
62. Lampe, C., Ellison, N., Steinfield, C.: A Face(book) in the crowd: social searching vs. social browsing. In: Proceedings of the 2006 20th Anniversary Conference on Computer Supported Cooperative Work, pp. 167–170 (2006)
63. Leistner, M.: The Commission's vision for Europe's digital future: proposals for the Data Governance Act, the Digital Markets Act and the Digital Services Act—a critical primer. J. Intell. Prop. Law Pract. **16**, 780 (2021)
64. Lynskey, O.: Grappling with data power: normative nudges from data protection and privacy. Theor. Inq. L 189, 196 (2019)
65. Marwick, A.: The public domain: surveillance in everyday life. Surveill. Soc. **9**, 378–393 (2012)
66. Maurer, M., Behringer, R., Furst, S., Thomanek, F., Dickmanns, E.D.: A compact vision system for road vehicle guidance. In: 13th International Conference on Pattern Recognition (1996)
67. Mayer-Schönberger, V., Cukier, K.: Big Data: A Revolution That Will Transform How We Live, Work, and Think, pp. 173–174. First Mariner Books (2015)
68. Miyashita, H.: Human-centric data protection laws and policies: a lesson from Japan. Comput. Law Secur. Rev. **40**, 105487 (2021). https://doi.org/10.1016/j.clsr.2020.105487. Accessed 21 Feb 2025
69. Mulder, T., Vellinga, N.E.: Exploring data protection challenges of automated driving. Comput. Law Secur. Rev. **40**, 105530 (2021)
70. Naylor, B.: Targeting online privacy, Congress sets a new tone with big tech. NPR, 13 Mar 2019. https://www.npr.org/2019/03/13/702619020/targeting-online-privacy-congress-sets-a-new-tone-with-big-tech. Accessed 21 Feb 2025
71. Opinion 1/2008 on data protection issues related to search engines (WP148), section 4.1.3, p. 12
72. EDPB, 2019, Opinion 5/2019 on the interplay between the ePrivacy Directive and the GDPR, in particular regarding the competence, tasks and powers of data protection authorities. Adopted 12 Mar 2019. https://www.edpb.europa.eu/sites/default/files/files/file1/201905_edpb_opinion_eprivacydir_gdpr_interplay_en_0.pdf. Accessed 22 Apr 2025
73. Packin, N.G., Aretz, Y. L.: Algorithmic analysis of social behaviour for profiling, ranking, and assessment, 16 Sept 2019. Accessed 21 Feb 2025
74. Pagani, M., Pardo, C.: The impact of digital technology on relationships in a business network. Ind. Mark. Manage. **67**, 185–192 (2017). https://doi.org/10.1016/j.indmarman.2017.08.009
75. Pasquale, F.: The Black Box Society: The Secret Algorithms That Control Money and Information, p. 61. Harvard University Press (2015)

76. Petit, N.: The proposed digital markets act (DMA): a legal and policy review. J. Eur. Compet. Law Pract. **12**, 532 (2021)
77. Quelle, C.: Enhancing compliance under the general data protection regulation: the risky upshot of the accountability- and risk-based approach. Eur. J. Risk Regul. **9**, 502, 518 (2018)
78. Ram, P., Markkula, J., Friman, V., Raz, A.: Security and privacy concerns in connected cars: a systematic mapping study. In: 2018 44th Euromicro Conference on Software Engineering and Advanced Applications (SEAA). IEEE (2018)
79. Regulation (EU) 2016/679 of the European Parliament and of the Council of 27 April 2016 on the Protection of Natural Persons with Regard to the Processing of Personal Data and on the Free Movement of Such Data, and Repealing Directive 95/46/EC (General Data Protection Regulation) (Text with EEA Relevance), OJ L 119, 4.5.2016, pp. 1–88
80. Regulation (EU) 2018/858 of the European Parliament and of the Council of 30 May 2018 on the Approval and Market Surveillance of Motor Vehicles and Their Trailers, and of Systems, Components and Separate Technical Units Intended for Such Vehicles, Amending Regulations (EC) No 715/2007 and (EC) No 595/2009 and Repealing Directive 2007/46/EC (Text with EEA Relevance), PE/73/2017/REV/1, OJ L 151, 14.6.2018, pp. 1–218
81. Regulation (EU) 2019/2144 of the European Parliament and of the Council of 27 November 2019 on Type-Approval Requirements for Motor Vehicles and Their Trailers, and Systems, Components and Separate Technical Units Intended for Such Vehicles, as Regards Their General Safety and the Protection of Vehicle Occupants and Vulnerable Road Users, Amending Regulation (EU) 2018/858 of the European Parliament and of the Council and Repealing Regulations (EC) No 78/2009, (EC) No 79/2009 and (EC) No 661/2009 of the European Parliament and of the Council and Commission Regulations (EC) No 631/2009, (EU) No 406/2010, (EU) No 672/2010, (EU) No 1003/2010, (EU) No 1005/2010, (EU) No 1008/2010, (EU) No 1009/2010, (EU) No 19/2011, (EU) No 109/2011, (EU) No 458/2011, (EU) No 65/2012, (EU) No 130/2012, (EU) No 347/2012, (EU) No 351/2012, (EU) No 1230/2012 and (EU) 2015/166 (Text with EEA Relevance), PE/82/2019/REV/1, OJ L 325, 16.12.2019, pp. 1–40
82. Regulation (EU) 2022/868 of the European Parliament and of the Council of 30 May 2022 on European Data Governance and Amending Regulation (EU) 2018/1724 (Data Governance Act) (Text with EEA Relevance), PE/85/2021/REV/1, OJ L 152, 3.6.2022, pp. 1–4
83. Regulation (EU) 2022/1925 of the European Parliament and of the Council of 14 September 2022 on Contestable and Fair Markets in the Digital Sector and Amending Directives (EU) 2019/1937 and (EU) 2020/1828 (Digital Markets Act) (Text with EEA Relevance), PE/17/2022/REV/1, OJ L 265, 12.10.2022, pp. 1–66
84. Regulation (EU) 2022/2065 of the European Parliament and of the Council of 19 October 2022 on a Single Market for Digital Services and Amending Directive 2000/31/EC (Digital Services Act) (Text with EEA Relevance), PE/30/2022/REV/1, OJ L 277, 27.10.2022, pp. 1–102
85. Regulation (EU) 2023/2854 of the European Parliament and of the Council of 13 December 2023 on Harmonised Rules on Fair Access to and Use of Data and Amending Regulation (EU) 2017/2394 and Directive (EU) 2020/1828 (Data Act) (Text with EEA Relevance), PE/49/2023/REV/1, OJ L, 2023/2854, 22.12.2023
86. Regulation (EU) 2024/1689 of the European Parliament and of the Council of 13 June 2024 Laying Down Harmonised Rules on Artificial Intelligence and Amending Regulations (EC) No 300/2008, (EU) No 167/2013, (EU) No 168/2013, (EU) 2018/858, (EU) 2018/1139 and (EU) 2019/2144 and Directives 2014/90/EU, (EU) 2016/797 and (EU) 2020/1828 (Artificial Intelligence Act) (Text with EEA Relevance), E/24/2024/REV/1, OJ L, 2024/1689, 12.7.2024, ELI: http://www.data.europa.eu/eli/reg/2024/1689/oj. Accessed 21 Feb 2025
87. Ren, Q., Du, J.: Harmonizing innovation and regulation: the EU artificial intelligence act in the international trade context. Comput. Law Secur. Rev. **54**, 106028 (2024)
88. Robertson, V.H.S.E.: Excessive data collection: privacy considerations and abuse of dominance in the era of big data. Common Mark. Law Rev. **57**, 161, 163 (2020)
89. Safjan, M.: Prawo do ochrony życia prywatnego i autonomia jednostki w orzecznictwie Trybunału Konstytucyjnego. In: Zubik, M. (ed.) Księga XX-lecia orzecznictwa Trybunału Konstytucyjnego. Warsaw (2006)

90. Sahebi, S., Formosa, P.: Social media and its negative impacts on autonomy. Philos. Technol. **35**, 70 (2022)
91. Salvini, P., Kunze, L., Jirotka, M.: On self-driving cars and its (broken?) promises. A case study analysis of the German Act on autonomous driving. Technol. Soc. **78**, 102628 (2024)
92. Schütz-Bosbach, S., Prinz, W.: Mirrors match minds. In: Ferrari, P.F., Rizzolatti, G. (eds.) New Frontiers in Mirror Neuron Research, pp. 198–221. Oxford University Press, Oxford, UK (2015)
93. Scott, J.D.: Social media and government surveillance: the case for better privacy protections for our newest public space essay. J. Bus. Technol. Law **12**, 151 (2016)
94. Tbatou, S., Ramrami, A., Tabii, Y.: Security of communications in connected cars modelling and safety assessment. In: Proceedings of the 2nd International Conference on Big Data, Cloud and Applications. BDCA'17, pp. 1–7 (2017)
95. The Constitution of Japan, Promulgated on 3 November 1946. https://www.japan.kantei.go.jp/constitution_and_government_of_japan/constitution_e.html. Accessed 24 Feb 2025
96. The Constitution of the Republic of Poland of 2 April 1997 (Journal of Laws No. 78, Item 483, as Amended)
97. Thrun, S., Montemerlo, M., Dahlkamp, H., Stavens, D., Aron, A., Diebel, J., Fong, P., Gale, J., Halpenny, M., Hoffmann, G., Lau, K., Oakley, C., Palatucci, M., Pratt, V., Stang, P., Strohband, S., Dupont, C., Jendrossek, L., Koelen, C., Markey, C., Rummel, C., van Niekerk, J., Jensen, E., Alessandrini, P., Bradski, G., Davies, B., Ettinger, S., Kaehler, A., Nefian, A., Mahoney, P.: The robot that won the DARPA grand challenge. J. Field Robot. **23**(9), 661–662 (2006)
98. Turienzo, J., Cabanelas, P., Lampon, J.F.: Business models in times of disruption: the connected and autonomous vehicles (uncertain) domino effect. J. Bus. Res. **156**, 113481 (2023). https://doi.org/10.1016/j.jbusres.2022.113481
99. Tuxera: Autonomous cars generate more than 300 TB of data per year, 2 July 2021. https://www.tuxera.com/blog/autonomous-cars-300-tb-of-data-per-year/. Accessed 21 Jan 2025
100. Valdez, J.: The politics of Uber: infrastructural power in the United States and Europe. Regul. Gov. **17**(1), 177–194 (2023)
101. Van Alsenoy, B.: Liability under EU data protection law: from directive 95/46 to the general data protection regulation. J. Intell. Prop. Inf. Technol. Electron. Commer. Law **7**, 271, para 43 (2016)
102. van Dijck, J., Nieborg, D., Poell, T.: Reframing platform power. Internet Policy Rev. **8** (2019)
103. Warren, S.D., Brandeis, L.D.: The right to privacy. Harvard Law Rev. **IV**(5) (1890)
104. Wasastjerna, M.: Competition, data and privacy in the digital economy: towards a privacy dimension in competition policy? Kluwer Law Int. 85–90 (2020)
105. Whitaker, R.: The end of privacy: how total surveillance is becoming a reality. In: Campbell, J.E., Carlson, M. (eds.) Panopticon.com: Online Surveillance and the Commodification of Privacy. J. Broadcast. Electron. Media **46**, 598 (2002)
106. Yu, S.H., Lindenberg, P.P., Cheng, B.C., Chen, H.: SMILE+: an efficient privacy preserving missed connection service in IoV networks. In: Park, J.H., Pan, Y., Kim, C., Yang, Y. (eds.) Future Information Technology—II. Springer Netherlands, Dordrecht (2015)
107. Zuboff, S.: The Age of Surveillance Capitalism: The Fight for a Human Future at the New Frontier of Power, p. 192. Profile Books (2019)
108. A. Theodorou, V. Dignum, Towards ethical and socio-legal governance in AI, Nature Machine Intelligence, 2 (1) (2020), pp. 10-12.

Chapter 8
Cybersecurity in Digital Mobility Services Within the Framework of Autonomous Vehicle Application

8.1 Introduction

Ensuring cybersecurity is one of the critical aspects of implementing digital mobility services. The significance of this issue is evidenced by statistical data from recent years regarding security incidents in IT infrastructures. According to Eurostat [16], 22% of EU companies experienced various repercussions due to ICT-related security incidents in 2021. These repercussions include service unavailability, data destruction or corruption, and even the disclosure of confidential information. Although a considerable number of incidents can be attributed to hardware or software failures, the frequency of malicious incidents is also rising. Ransomware attacks increased by 41% in 2022 [17].

In a more confined context, the matter of cybersecurity concerns the assurance of the technical feasibility of establishing secure communication between the parties engaged in contracts related to the provision of digital mobility services and the utilisation of autonomous vehicles. In this regard, cybersecurity has implications for both the viability and the financial sustainability of delivering such services. A study conducted by the Joint Research Centre in 2020 projected that the global cost of cybercrime would reach EUR 5.5 trillion by the end of 2020, an increase from EUR 2.7 trillion in 2015 [14]. Projections for 2025 indicate that figures may escalate to USD 10.5 trillion [27].

The justification for providing digital mobility services requires, among other things, the premise of the security of autonomous vehicles in road traffic. This premise necessitates a more comprehensive analysis of cybersecurity.

A. Krasuski, *Legal Challenges of a Transformation from a Product-Centric to a Service-Centric Business Model and Procedures*, Studies in Systems, Decision and Control 610, https://doi.org/10.1007/978-3-031-98901-8_8

8.2 Rationale for Cybersecurity Regulation

The transition of automotive systems from electromechanical components to those that are electronically and software-controlled has facilitated real-time connectivity for autonomous vehicles. This evolution has concurrently integrated technology companies, software developers, and sensor manufacturers into the operations of autonomous vehicles [23].

According to Vandezande [42], the European Commission acknowledged the rapidly increasing reliance of society on an ever more interconnected network and information systems, as outlined in the EU Cybersecurity Strategy 2020. This is evidenced by the rising number of connected devices [40], including autonomous vehicles. The authorisation of autonomous vehicles for use in road traffic, along with the provision of digital mobility services linked to these vehicles, requires the processing of substantial volumes of data through ICT networks.

Dukarski [13] asserts that modern automobiles accumulate approximately 25 GB of data per hour, with autonomous vehicles anticipated to generate over 100 times that amount. In addition to the volume of data, the methodology of its collection also suggests a vast range of individuals to whom this data is relevant. Utilising sensors and cameras integrated within an autonomous vehicle, data is collected not only from the current driver and passengers but also from other road users and bystanders. According to Miller et al. [26], tracking data, which measures body posture and movement, serves as a significant source of information. Such data facilitates associations between various forms of bodily motion and outcomes related to creativity, learning, and other behavioural results. Currently, there exists no specific legal framework or formal consent pertaining to the monitoring or dissemination of bystanders' data. Consequently, the repercussions of cybercrime extend beyond the directly affected entities. Given that over 45% of breaches involve personal data, this situation exposes citizens globally to a myriad of risks, including identity theft and financial fraud [45].

Khan et al. [22] appropriately contend that a successful cyberattack on autonomous vehicles results in breaches of both privacy and security. Regarding security implications, the threats include system failures, inaccurate speed and location assessments, failures in object detection, traffic congestion, misleading toxic signage, and compromised longitudinal safety and vehicle control.

Autonomous vehicles, integrated with various components of road infrastructure, have the capacity to navigate independently and analyse substantial volumes of data. It is imperative to distinguish the information qualifying as personal data, as defined in Article 4 of Regulation (EU) 2016/679 [30]. Member States of the European Union may enact provisions within their legal frameworks that clearly identify the individual whose personal data is being processed. The scope of the personal data being processed may also include health information related to the individual acting as the actual driver. The justification for processing such data, including personal data, stems from the unique nature of the movement of connected and autonomous vehicles, their interaction with other components of road infrastructure, and the involvement of individuals. The exchange of information between autonomous vehicles and infrastructure elements occurs through radio networks. Without this capacity for data exchange, the operation of vehicles characterised by such intricate data interactions would not be feasible. Consequently, it can be inferred that the security of data, including personal data processed via mobile networks, is vulnerable to various cybersecurity threats.

Given the complexity surrounding the coexistence of the aforementioned data processing sources, it is imperative to ensure cybersecurity for the following aspects: (1) the security of processing personal data relating to the human driver and passengers; (2) the security of processing personal data of other road users; and (3) the assurance of security in processing data of individuals who are not road users, whose information is collected in relation to scanning the topography of densely populated areas, such as urban centres.

8.3 The Concept of Cybersecurity

The concept of 'cyber security' has been articulated in the literature with an emphasis on technical aspects, while the socio-technical dimensions related to the application of technology have often been overlooked. It is understood that this concept encompasses the strategies and measures employed by organisations and governments to address security threats, ensuring the confidentiality, integrity, and availability of data and digital assets within cyberspace [38]. The International Telecommunication Union (ITU) [18], in its recommendations, proposes the following definition of cybersecurity: it is described as a comprehensive framework comprising tools, policies, security concepts, safeguards, guidelines, risk management strategies, activities, training, best practices, and technologies utilised to protect the cyber environment and assets of both organisations and users. The interpretation of 'organisation and user assets' is extensive, encompassing connected computing devices, personnel, infrastructure, applications, services, telecommunications systems, and all information transmitted and/or stored within the cyber environment. Consequently, cybersecurity aims to ensure that the security properties of organisational and user assets are achieved and maintained against pertinent security threats in the cyber environment. General security objectives encompass availability, integrity (which may include authenticity and non-repudiation), and confidentiality.

In the existing legislative framework, the concept of 'cybersecurity' is defined in Regulation (EU) 2019/881 ('Cybersecurity Act') [31] and encompasses the activities essential for safeguarding network and information systems, the users of these systems, and other individuals affected by cyber threats (Article 2(1)).

In comparison to the existing Directive (EU) 2022/2555 ('NIS2 Directive') [9], the previously enacted Directive 2016/1148 ('NIS Directive') [7] primarily focused on the 'security of network and information systems'. The NIS2 Directive, however, encompasses a more comprehensive concept of 'cybersecurity', as delineated in

Article 2(1) of the Cybersecurity Act. This adjustment signifies that the objective extends beyond the mere protection of network and information systems to include 'the users of such systems, and other persons affected by cyber threats'. Given the significant risks that cyberattacks present to users of ICT systems, this broader focus represents a valuable advancement.

Furthermore, the term 'network and information system' must be interpreted as outlined in point (1) of Article 6 of the NIS2 Directive. This definition includes:

(1) an electronic communications network as defined in point (a) of Article 2, point (1) of Directive (EU) 2018/1972 [8];
(2) any device or group of interconnected or related devices, one or more of which, under a programme, engage in the automatic processing of digital data; or
(3) digital data that is stored, processed, retrieved, or transmitted by elements detailed in points (a) and (b) for their operation, use, protection, and maintenance.

Within the European Union, there is an increasing prevalence of devices connected to the internet. However, security measures and resilience protocols, particularly in the aftermath of a security breach, are not adequately integrated by design. This shortfall leads to insufficient cybersecurity.

8.4 General Legal Framework on Cybersecurity

The exponential increase in cybersecurity challenges in recent years, particularly the surge in malware attacks, has necessitated the development of new mandatory regulations by European Union institutions. Malware, defined as malicious software, refers to any programme or file that is specifically crafted to harm or exploit programmable devices or networks. It serves as a comprehensive term encompassing various harmful programmes, including but not limited to computer viruses, worms, Trojan horses, and ransomware, which are deployed to inflict damage or gain unauthorised access to sensitive information. At the institutional level within the European Union, the issue of cybersecurity was previously addressed through the adoption of the Cybersecurity Strategy of the European Union in 2013, aimed at guiding the Union's policy response to cyber threats and risks. To further bolster the protection of citizens online, the Union established its inaugural legal framework in the realm of cybersecurity in 2016 with the adoption of Directive 2016/1148, commonly referred to as the 'NIS Directive'. This directive delineated significant milestones across various sectors:

(1) establishing requirements concerning national capabilities in the field of cybersecurity;
(2) setting forth the first mechanisms to enhance strategic and operational cooperation between Member States; and
(3) introducing obligations concerning security measures and incident notifications across sectors that are vital for the economy and society, such as energy, transport, drinking water supply and distribution, banking, financial market

infrastructures, healthcare, digital infrastructure, as well as key digital service providers.

On 16 January 2023, the NIS2 Directive came into force, superseding Directive (EU) 2016/1148. The NIS2 Directive takes on the status of lex generalis among the various cybersecurity provisions that make up the EU legal framework for cybersecurity. Despite the perceived divergence in the interpretation of the NIS Directive across the Member States and the resulting discrepancies in its implementation within the national laws of the Member States [15], the EU legislator did not choose to introduce a new regulation through an EU regulation.

The NIS2 Directive concerns entities operating in the sectors listed in the annexes of the Directive, including providers of data centre and cloud computing services, as well as providers of public electronic communication networks. These entities must satisfy specific company size criteria and must function or deliver their services within the European Union. This Directive highlights the need for a governance framework that outlines the roles and responsibilities of relevant stakeholders at the national level, thereby cataloguing the various authorities and stakeholders involved in implementing the strategy, while also establishing a policy framework designed to enhance coordination among the appropriate authorities (Article 7(1) of the NIS Directive). Furthermore, the NIS2 Directive regulates the Cooperation Group (Article 14) and the computer security incident response teams (CSIRTS) (Article 15). In accordance with Article 16, the European Cyber Crisis Liaison Organisation Network (EU-CyCLONe) is introduced as a new entity.

In accordance with Article 20 of the NIS2 Directive, the management bodies of entities are required to formally authorise the risk management measures relevant to their organisations and oversee the implementation of these measures. They may be held accountable for any violations pertaining to this framework. Moreover, they are obliged to engage in relevant training to ensure they possess adequate knowledge and skills.

A significant issue concerning the assurance of cybersecurity in the deployment of autonomous vehicles, particularly regarding the provision of digital mobility services, is the recognition of the potential risks associated with cybersecurity breaches. Road safety now encompasses a new area of interest: the cybersecurity of autonomous vehicles. Within this context, human vulnerability emerges as the most likely catalyst for a successful cyberattack. To mitigate this challenge, scholars specialising in human factors and psychology can play a pivotal role in enhancing the cybersecurity of autonomous vehicles by exploring methodologies to reduce the likelihood of such attacks. This investigation would adopt a human-centred perspective, considering the adversary's capacity to select the mode of attack. The two primary elements that facilitate awareness are the users of autonomous vehicles and the effectiveness of educational initiatives in this domain [24]. As stated by Vandezande [42], it is estimated that 95% of cybersecurity issues are attributable to human error, thereby underscoring the imperative to elevate public cognition regarding cybersecurity.

The NIS2 Directive references various legal instruments, including, among others, the Cybersecurity Act. This regulation addresses two critical aspects. First, it establishes an EU-wide certification scheme for ICT products, processes, and services, which is developed and maintained by the European Union Agency for Network and Information Security (ENISA), as delineated in Regulation (EU) No 526/2013 [36]. This certification allows for a singular certification of these products, processes, and services across the entire EU, thereby providing users with assurance regarding their cyber resilience. Moreover, the certification scheme is incorporated within the NIS2 Directive, enabling Member States to mandate the utilisation of certified products, processes, and services in specific instances.

Burzio et al. [2] presented a ranking-based framework for the high-level certification of cybersecurity in autonomous vehicles, utilising two principal methodologies: (1) vehicle security in accordance with the Euro New Car Assessment Program evaluation, and (2) a proposed cybersecurity ranking methodology that involves the assessment of components, subsystems, and connection topology to determine the ranking of the overall system. However, the programme primarily addresses the overarching hierarchy within the domain of automotive cybersecurity and lacks comprehensive methodologies to thoroughly investigate this issue [23]. The concept of establishing a security ranking is based on the methodology already incorporated within the Euro NCAP assessments for vehicle safety evaluation.

Raiyn [29] proposed a cyberattack defence paradigm that utilises iris recognition, integrating biometric data for message authentication and secure transmission. However, the study predominantly focuses on flowcharts and relies on a singular analytical stream—communication among autonomous vehicles.

According to Article 1(2) of the NIS2 Directive, Member States are obligated to 'adopt national cybersecurity strategies and to designate or establish competent authorities, cyber crisis management authorities, single points of contact on cybersecurity (single points of contact), and CSIRTS'. The NIS2 Directive introduced measures for cybersecurity risk management and reporting, alongside regulations and obligations concerning the sharing of cybersecurity information, as well as supervisory and enforcement responsibilities for Member States. As noted by Vandezande [42], these enhancements exemplify the NIS2's increased emphasis on the effective enforcement of its measures, which, as demonstrated by the aforementioned impact assessment, were not thoroughly implemented under the NIS Directive. Entities governed by the NIS2 Directive are required to comply with specific cybersecurity mandates, encompassing requirements related to cybersecurity risk governance, management, and reporting obligations. The Directive adopts a minimum harmonisation approach, allowing Member States to enact provisions that ensure a higher level of cybersecurity. It is essential for businesses to carefully monitor national implementations to ensure compliance with the relevant provisions.

The NIS2 Directive applies to entities engaged in the sectors listed in the annexes of the Directive, which include data centre and cloud computing service providers, as well as providers of public electronic communications networks. Such entities must meet the specified company size criteria and must operate or provide their services within the European Union. Moreover, this category includes domain name registration service providers and those classified as critical under Directive (EU) 2022/2557, referred to as the 'CER Directive' [10].

The NIS2 Directive applies, regardless of size, to entities classified as belonging to the categories specified in Annex I (sectors of high criticality) or Annex II (other critical sectors). This includes the following conditions:

(1) The services are provided by providers of public electronic communications networks or publicly available electronic communications services; trusted service providers; top-level domain name registries; and domain name system service providers.
(2) The entity acts as the sole provider within a Member State of a service that is essential for maintaining critical societal or economic activities.
(3) Disruption of the service provided by the entity could significantly impact public safety, public security, or public health.
(4) Disruption of the service offered by the entity could introduce a substantial systemic risk, particularly in sectors where such disruption may have cross-border implications.
(5) The entity is considered critical due to its specific importance at the national or regional level concerning the particular sector or type of service, or for other interdependent sectors within the Member State.
(6) The entity is identified as a public administration entity, whether at the central or regional level.

While the NIS Directive categorises digital service providers distinctly from essential entities, certain providers in this domain are considered highly critical under the NIS2 Directive. Notable inclusions within this classification comprise, among other entities, Internet exchange points ('IXP'), which serve as physical locations facilitating connectivity among Internet infrastructure companies, including Internet Service Providers ('ISPs') and Content Delivery Networks ('CDNs') [4]. This classification further encompasses cloud services, data centres, content delivery services, trust service providers, and public electronic communications services and networks. Cloud services, previously regarded as digital services under the NIS Directive, have been elevated to a higher tier under the NIS2 Directive.

Within NIS2, additional significant sectors include, among others, computers, electronic and optical products, electrical equipment, machinery, motor vehicles and trailers, and other transport equipment. Furthermore, the category of digital providers comprises online marketplaces, online search engines, and social networking platforms.

The CER Directive designates specific entities across eleven sectors as critical entities. This classification largely coincides with the designation of entities by the NIS2 Directive as essential entities, a decision made intentionally. The CER Directive primarily addresses the physical security of such entities, while the NIS2 Directive pertains to their cybersecurity. Given the interconnectedness of these aspects, entities designated under both directives are required to adhere to both regulatory frameworks. Furthermore, competent authorities within both frameworks are encouraged to collaborate closely on relevant issues. The CER Directive emphasises enhancing physical resilience among critical entities providing essential services across eleven pivotal sectors, including energy, transport, and healthcare. This Directive became effective in January 2023.

Another initiative is Regulation (EU) 2022/2554 [33] concerning digital operational resilience for the financial sector (hereinafter referred to as the 'DORA Act'). The DORA Act, an EU regulation, came into effect on 16 January 2023 and will be applicable from 17 January 2025. This regulation is designed to enhance the IT security of financial entities such as banks, insurance companies, and investment firms while ensuring that the financial sector in Europe retains its resilience in the face of significant operational disruptions. The DORA Act introduces harmonisation of the rules governing the operational resilience of the financial sector, applicable to twenty different categories of financial entities and third-party ICT service providers. DORA places a specific emphasis on financial entities, listing twenty-one types of regulated entities, including credit institutions, payment and e-money institutions, investment firms, insurance and reinsurance undertakings, and crowdfunding service providers, among others. In addition to the financial entities themselves, DORA also extends its application, to some degree, to third parties providing ICT-related services to them. The framework seeks to establish common requirements for their digital operational resilience, which inherently includes cybersecurity.

In 2022, the European Commission implemented an enhancement to the EU cybersecurity framework, known as Regulation (EU) 2024/2847 [35], referred to as the 'Cyber Resilience Act' or 'CRA'. This regulation seeks to 'strengthen the Union's approach to cybersecurity, address cyber resilience at Union level and improve the functioning of the internal market by laying down a uniform legal framework for essential cybersecurity requirements for placing products with digital elements on the Union market' (Recital 1 of the CRA).

This framework delineates the cybersecurity requirements applicable to products—both hardware and software—that incorporate digital components, as specified in Article 2(1) and Article 3(1) of the CRA. Its primary objective is to address the challenges associated with devices, including computers and smartphones, which are often introduced into the marketplace with inherent security vulnerabilities or a lack of security updates throughout their operational lifespan. Although the NIS2 framework pertains to specific service providers, such as those engaged in cloud computing, it does not comprehensively cover all manufacturers and suppliers of hardware and software that are vulnerable to security threats. The legislation enumerates a definitive list of 'important products with digital elements' in Annex III, along with a clear catalogue of 'critical products with digital elements' in Annex IV, thus providing legal certainty.

The Cyber Resilience Act ('CRA') addresses a distinct cybersecurity concern that is not comprehensively covered by the NIS2 framework. The CRA mandates cybersecurity requirements for products featuring digital elements, requiring manufacturers and retailers to adhere to established security standards. The CRA applies to interconnected software and hardware products, with specific exceptions for items governed by particular legislation. Products incorporating digital elements may be introduced into the European market, provided that, pursuant to Article 6(a) and (b) of the CRA, they comply with the essential requirements outlined by law, specifically regarding product characteristics (refer to Annex I, Part I of the CRA) and requirements for vulnerability management (refer to Annex I, Part II of the CRA).

In accordance with Article 13(12) of the CRA, manufacturers are required to carry out the designated conformity assessment procedures as outlined in Article 32, or to ensure that such procedures are carried out. The conformity procedures are detailed in Annex VIII. These procedures include internal controls executed by the manufacturer, as well as an evaluation conducted by a third party appointed by a national authority.

The CRA introduces a general requirement in Article 6(a) (b) that products, prior to being introduced on the market, must: (a) meet the essential cybersecurity requirements set out in Part I of Annex I, provided that they are correctly installed, maintained, and used for their intended purpose or under conditions that can reasonably be foreseen, and, where applicable, the necessary security updates have been installed; and (b) ensure that the processes implemented by the manufacturer comply with the essential cybersecurity requirements set out in Part II of Annex I.

Manufacturers are obligated to ensure the effective management of vulnerabilities throughout the anticipated lifespan of the product or for a minimum duration of five years, whichever period is shorter, subsequent to the sale of the product (Article 13(8); Annex I, Part II of the CRA). They are required to identify and meticulously document dependencies and vulnerabilities, including a software bill of materials (Annex I, Part II(1) of the CRA). In accordance with Article 3(39) of the CRA, a 'software bill of materials' is defined as a formal record that encompasses details and supply chain relationships of components integrated within the software elements of a product with digital characteristics.

As articulated by Colonna [5], manufacturers must promptly address identified vulnerabilities, conduct comprehensive security testing of the digital product, publicly disclose information regarding resolved vulnerabilities, establish a coordinated vulnerability disclosure policy, facilitate the exchange of information about potential vulnerabilities, provide mechanisms for secure updates, and expeditiously deliver patches at no cost, accompanied by advisory messages (Annex I, Part II of the CRA).

Article 14(1) of the CRA outlines the manufacturers' obligation to notify any actively exploited vulnerabilities in products with digital elements of which they become aware. This notification must be made simultaneously to the CSIRT designated as the coordinator, as specified in paragraph 7 of this article, and to ENISA. Furthermore, manufacturers are required to report any actively exploited vulnerabilities through the single reporting platform. Additionally, there is a duty to inform the individual or entity responsible for maintaining the component integrated into the product if it is affected by a vulnerability identified by the manufacturer, as stated in Article 13(6) of the CRA.

Where, in the course of that evaluation, the market surveillance authority finds that a product with digital elements does not comply with the requirements laid down in the Cyber Resilience Act, it shall without delay require the relevant economic operator to take all appropriate corrective actions to bring the product with digital elements into compliance with those requirements, to withdraw it from the market, or to recall it within a reasonable period, commensurate with the nature of the cybersecurity risk, as prescribed by the market surveillance authority (Article 54(1) of the CRA). They also have the authority to impose restrictions on the availability of a product within the European market or to mandate product recalls (Article 54(5) of the CRA).

Should the market surveillance authority determine during the assessment that a product containing digital elements fails to comply with the requirements set out in the Cyber Resilience Act, it must promptly require the relevant economic operator to undertake all necessary corrective actions to ensure compliance, withdraw the product from the market, or initiate a recall within a reasonable timeframe appropriate to the nature of the cybersecurity risk. This action is consistent with the provisions laid out by the market surveillance authority (Article 54(1) of the Cybersecurity Act). Moreover, the market surveillance authorities have the power to impose restrictions on the availability of the product within the European market or to mandate its withdrawal (Article 54(5) of the CRA).

The Corporate Responsibility Act will come into effect on 10 December 2024, with its primary provisions scheduled for implementation commencing on 11 December 2027. The reporting obligations for producers are set to begin on 11 September 2026.

8.5 Cybersecurity Obligations Under the Legal Framework for Product Liability

8.5.1 General Remarks

On 28 September 2022, the European Commission published its proposal for a Directive concerning Liability for Defective Products, thereby updating the nearly 40-year-old Product Liability Directive (Directive 85/374/EEC) [6]. After extensive negotiations, this proposal was officially released in the Official Journal of the European Union on 18 November 2024 (Directive (EU) 2024/2853, 'Product Liability Directive' or 'PLD', 2024) [11] and came into effect on the twentieth day following its publication (Article 23 of the PLD). The legislation seeks to modernise the product liability framework within the European Union to effectively tackle the challenges posed by the digital age and simplify the process for consumers seeking compensation for damages caused by defective products.

In relation to the damages caused by defective products, the Product Liability Directive establishes a no-fault liability regime applicable to manufacturers. As outlined in Article 21(1) of the PLD, Directive 85/374/EEC will remain in force and is set to be repealed on December 9, 2026. However, this directive will still apply to products that were placed on the market or put into service before this date.

Furthermore, according to Article 22(1) of the PLD, Member States are required to implement the provisions of the Product Liability Directive by December 9, 2026.

8.5.2 Analysing the Digital Mobility Services in Accordance with Directive 85/374/EEC

8.5.2.1 Software as a Product Under Directive 85/374/EEC

One of the objectives of the Product Liability Directive ('PLD') is to adapt the definition of 'product' to the evolving manner in which it is utilised within the digital realm. Article 2 of the currently binding Directive 85/374/EEC defines the term 'product' by referencing the concept of 'movable things'. Consequently, products are categorised as movable entities, exemplified by items such as mobile phones. Furthermore, Article 2 of the PLD explicitly includes electricity within its scope. The classification of software as a 'product' generates uncertainties within legal discourse [1]. Software lacks tangibility and does not constitute a movable entity, unless it is stored on a tangible medium, such as a hard drive or USB flash drive. Accordingly, over-the-air updates would fall outside the jurisdiction of Directive 85/374/EEC [44]. In its judgment dated 10 June 2021, case C-65/20 [20], the European Court of Justice determined that Article 2 of Directive 85/374/EEC should be interpreted to mean that a printed copy of a newspaper which, in the context of paramedical matters, contains misleading health advice regarding the use of a particular plant—subsequently proven harmful to the health of the reader—does not qualify as a 'defective product' under the provisions of Directive 85/374/EEC. The court elucidated in recital 27 of this ruling that 'services do not fall within the scope of that directive'. Additionally, according to Wolters [48], embedded software is considered an integral component of a product. Within this framework, the term 'product' encompasses all forms of software, provided it is installed within the product.

8.5.2.2 The Right to Pursue Compensation for Damages in Accordance with Directive 85/374/EEC

According to Article 1 of Directive 85/374/EEC, the producer is held liable for damages resulting from a defect in the product. The term 'producer' is defined as the manufacturer of the finished product, the producer of the raw material, or the producer of a component, as well as any individual who, by affixing their name, trademark, or any other distinguishing feature to the product, presents themselves as the producer (Article 3(1)). Furthermore, the legislator classifies any individual who imports a product into the European Union for purposes of sale, hire, leasing, or any form of distribution within the scope of their business as a producer (Article 3(2)). In instances where a product has multiple manufacturers (for example, in the case of automated vehicles), these manufacturers bear joint and several liability for the damage incurred (Article 5 of Directive 85/374/EEC).

Article 9 of Directive 85/374/EEC delineates two categories of damages for which compensation may be sought from the producer. The first category encompasses damages resulting in death or personal injury. The second category pertains to material damages, which also includes the destruction of property that is not the defective product itself, subject to the following two conditions: (1) the damage must be of a type ordinarily intended for private use or consumption, and (2) it must have been utilised by the injured party primarily for their private use or consumption. In accordance with Article 4 of Directive 85/374/EEC, claims for compensation are contingent upon the substantiation of damage, the identification of the defect, and the establishment of a causal relationship between the defect and the damage. The prevailing legal framework endorses a no-fault liability principle, which is currently confined to tangible products.

According to Article 6(1) of Directive 85/374/EEC, a product is deemed defective when it fails to provide the safety that an individual is entitled to expect, taking into account all relevant circumstances, including: (a) the presentation of the product; (b) the intended use for which the product could reasonably be expected; (c) the time at which the product was placed into circulation. Furthermore, Article 6(2) of Directive 85/374/EEC states that a product shall not be regarded as defective solely due to the introduction of a superior product into circulation subsequent to its own.

Furthermore, the regulations pertaining to liability for fault or negligence differ across various jurisdictions, thereby providing disparate levels of protection to victims. Nonetheless, Directive 85/374/EEC ensures a uniform level of protection for victims throughout the European Union. Consequently, the fault-based approach is considerably less favourable for victims. Claimants are required to comprehensively demonstrate that the product was defective and that this defect was the cause of the damage.

If software updates are not classified as a ‘product’ within the framework of Article 2 of Directive 85/374/EEC, several issues must be addressed. The injured party may seek compensation from the vehicle manufacturer as a whole, typically represented by the vehicle brand. In cases where an error in a software update has caused an accident, the manufacturer is likely to argue that the product—the vehicle—was not defective when it was released into circulation. This defence is outlined in Article 7(b) of Directive 85/374/EEC. This provision indicates that, under the circumstances, it is probable that the defect responsible for the damage did not exist at the time the product was placed into circulation by the manufacturer or that the defect manifested afterwards. According to Vellinga [44], it would be beneficial for manufacturers to distribute malicious software solely in the form of updates to mitigate liability under Directive 85/374/EEC. This scenario would create a distinction between the party harmed by an automated vehicle that was released into circulation with defective software pre-installed and the party harmed by an automated vehicle that obtained the defective software through an update post-circulation. Conversely, a manufacturer retains the ability to continuously modify its product through software updates.

Given that Directive 85/374/EEC serves as a legal instrument of the European Union, the Court of Justice of the European Union shall ultimately determine the interpretation of the concept of 'product' and, consequently, address the question regarding the circumstances under which a software product (update) may be classified as a 'product' as defined in Article 2 of the Directive.

8.5.3 Cybersecurity Obligations Under Directive (EU) 2024/2853

8.5.3.1 Revised Definition of Product

In comparison to Directive 85/374/EEC, Directive (EU) 2024/2853 broadens the definition of 'product' to encompass software, thereby challenging the conventional distinctions between products and services. Article 4(1) of the PLD elucidates that 'product' refers to 'any movable thing, whether or not integrated with or connected to another movable or immovable thing; this includes electricity, digital production files, raw materials, and software'. It is noteworthy that the PLD does not provide a definition for the term 'software'. According to Colonna [5], this term, integrated into the definition of 'product', denotes software that may be either stand-alone or embedded, irrespective of the means of delivery or access (e.g., on a device, via the cloud, or as a service), and is now explicitly recognised within the definition of 'product'. The term 'software' is notably absent from the PLD itself. Citing recital 13, the European legislator provides specific examples that contribute to the interpretation of this concept, which include 'operating systems, firmware, computer programs, applications, or AI systems'. Colonna [5] posits that source code does not qualify as a 'product,' nor does free and open-source software that is developed outside the context of commercial activities. Furthermore, in light of Wagner [46], the classification of software as a product for the purposes of product liability is of diminished practical significance for end producers, such as assemblers, who market combinations of hardware and software. This includes, for instance, vehicles equipped with autonomous driving capabilities or even vehicles with what might be considered 'simple' engine control software.

There exists widespread consensus that the final manufacturer is accountable for defects in what is referred to as 'embedded software' [41, 46]. The PLD additionally includes two critical definitions, namely 'digital manufacturing file' and 'component'. As explicated in Article 4(2), 'digital manufacturing file' signifies a digital representation or template of a movable object that encompasses the functional information necessary for the production of a tangible artefact by enabling the automated control of machinery or tools. Recital 16 cites 'a computer-aided design file utilised to create a 3D printer' as an example of digital manufacturing files.

Furthermore, the PLD has established an expansive definition of 'components' in Article 4(4), which, among other aspects, includes a 'related service' that is integrated or connected with a product. The incorporation of digital services through Articles 4(3) and 4(4) yields several implications. Notably, in the context of digital devices, the notion of a digital service encompasses the provision of data, as indicated by Recital 17 of the PLD. The essential consideration is whether a digital service qualifies as a 'related service' for the purposes of interpreting Article 4(3) of the PLD. The definition of a related service is further elucidated in Article 4(4), stipulating that a service must be integrated or connected with a product by or under the control of the manufacturer. If these criteria are satisfied, the service qualifies as a 'component' of the final product. In Recital 17 of the PLD, the EU legislator enumerates examples of related services, which include: the continuous provision of traffic data in a navigation system, a health monitoring service utilising sensors in a physical product to track the user's physical activity or health indicators, a temperature control service regulating the temperature of a smart refrigerator, or a voice assistant service enabling the control of one or more products through voice commands. It is important to emphasise that the EU legislator, in Recital 17 of the PLD, asserts that a product reliant on Internet access services that fails to maintain security (including cybersecurity) in the event of a loss of connection may be deemed defective under this directive. Nonetheless, Internet access services ought not to be classified solely as related services, as they cannot be regarded as components of a product under the manufacturer's control. Thus, it would be unreasonable to hold manufacturers accountable for damages arising from deficiencies in Internet access services. According to Koch [25], this signifies a significant departure from the prior emphasis on tangible products, as it now extends liability to service providers who deliver ongoing support following the distribution of the product—a situation that was previously excluded under the former regime.

Considering the above analysis, the third, fourth, and fifth types of digital mobility service indicate software that will be connected to an autonomous vehicle, which is a product within the meaning of PLD and Directive 85/374/EEC.

8.5.3.2 The Concept of Product Defect

The Product Liability Directive expands the responsibilities of the manufacturer beyond the initial market release, particularly regarding products that undergo updates or improvements over time. The manufacturer's liability includes circumstances related to software updates and upgrades, especially when these elements fall within the manufacturer's oversight. The phrase 'manufacturer's control' pertains to software updates, upgrades, and modifications that occur after their initial market release.

Under Directive 85/374/EEC, the framework for liability was primarily established for tangible goods, encompassing several exclusions from liability. The Product Liability Directive incorporates these exclusions. Conversely, Article 11(2) of the directive stipulates that liability is not excluded for related services or software (including updates), even in circumstances where the defect is likely not to have existed at the time the product was placed on the market. This divergent approach, in comparison to tangible goods, perpetuates the concept of the defect test, according to which the manufacturer's liability extends beyond the product's initial market release, provided that the manufacturer maintains control over the product. In practice, this exclusion may create an obligation to update products to comply with the latest safety standards, where such an obligation is not already established under the EU technical regulatory framework.

Colonna [5] asserts that software developers, including providers of artificial agent systems, are classified as manufacturers and may incur liability for defects associated with updates, enhancements, or machine learning algorithms. This principle holds true even when the software is integrated into another product developed by a different manufacturer. Furthermore, individuals or entities that significantly modify a product in the post-market phase, beyond the control of the original manufacturer, are regarded as manufacturers under Article 8(2) of the PLD. This interpretation expands strict liability to encompass digital services that are integrated with or connected to other products.

The Product Liability Directive considerably expands the criteria for evaluating whether a product is defective. According to the PLD, a product is considered defective if it 'does not provide the safety that can be expected or required by EU or national law' (Article 7(1) of the PLD). The determination of a 'defect' necessitates the consideration of all the criteria enumerated by the legislator in Article 7(2) of the PLD, which comprises a closed list. In assessing the existence of a product defect, the EU legislator mandates that relevant product safety requirements be taken into account, including pertinent cybersecurity requirements related to safety (Article 7(2)(f) of the PLD). As stated in recital 32 of the preamble to the PLD, the EU legislator regards a product as defective if it possesses a cybersecurity vulnerability. An illustrative example of such a defect would be a product failing to meet established cybersecurity requirements. Additional considerations include, inter alia, the reasonably foreseeable use of the product (Article 7(2)(f) PLD) and the reasonably foreseeable effects of other products that may reasonably be anticipated to be used in conjunction with the product, particularly through interconnections (Article 7(2)(f) of the PLD). The capacity of artificial intelligence to 'learn' or evolve after implementation introduces a further layer of complexity in assessing product safety and liability. The Product Liability Directive ensures that liability extends to changes in a product over time, particularly in instances where these modifications compromise safety or performance. An artificial intelligence system may be deemed defective not only based on its initial performance but also in relation to its behaviour following software updates or its interaction with other interconnected devices. The reference point for this assessment should be the average consumer.

Article 10 of the Product Liability Directive (PLD) continues to require claimants to establish the product's defectiveness, the damages incurred, and the causal relationship between these elements. However, in contrast to prior regulations, the PLD now significantly reduces the burden of proof for the claimant. According to the revised provisions, a product is presumed defective if the defendant fails to present evidence, the claimant proves that the product does not comply with mandatory safety standards, or the claimant demonstrates that an obvious malfunction occurred during foreseeable use (Article 10(2)(a)–(c) of the PLD). Furthermore, a causal link is presumed if it has been established that the product is defective and that the resultant damages are of a nature typically associated with the identified defect (Article 10(3) of the PLD). Additionally, should claimants face substantial challenges in demonstrating defectiveness or causation due to technical or scientific complexities, they are required only to indicate that it is likely the product is defective or that a causal link exists (Article 10(4) of the PLD). Consequently, the burden then shifts to the defendant to refute these presumptions, thereby streamlining the compensation process for affected parties (Article 10(5) of the PLD).

8.5.3.3 Liability for Damage

In comparison to Directive 85/374/EEC, the Product Liability Directive expands the definition of 'damage' as stipulated in Article 6(1)(a) to encompass death or personal injury, as well as medically recognised mental damage. Furthermore, Article 6(1)(b) specifies damage to or destruction of property, with specific exclusions as follows:

(1) The defective product itself,
(2) Property that sustains damage due to a defective component integrated into or connected with the product by the manufacturer, or otherwise under the control of that manufacturer,
(3) Property utilised solely for professional purposes.

The grounds for compensation include the destruction or damage to data that is not used for professional purposes, as stipulated in Article 6(1)(c) of the PLD. In accordance with Article 6(2) of the PLD, compensation covers all material losses resulting from the aforementioned damage. The European Union legislator has empowered the Member States with the discretion to determine how to calculate compensation. Furthermore, Article 6(2) of the PLD may also include non-material losses, dependent on their occurrence as a result of fatal or personal injury [5].

From the perspective of evaluating the scope of liability for damages resulting from deficiencies in digital mobility services, referencing recital 24 of the PLD, the European Union legislator intended to exclude claims for pure economic loss, breaches of privacy, or instances of discrimination under this Directive. Wagner [46] argues that this position can be justified by the notion that incentives to contribute to open-source software projects should not be undermined due to the emergence of new liability risks.

The Product Liability Directive imposes substantial obligations on manufacturers within the European Union to guarantee the safety and reliability of their products, taking into account the criteria for acknowledging 'defectiveness'. A manufacturer's liability for a defective product extends to any damage incurred as a result of a defective component when it is integrated or connected to a product under the manufacturer's oversight. Pursuant to Article 8 of the PLD, manufacturers are classified as 'economic operators' and may be held liable for defects in their products that cause harm to consumers. Consequently, the PLD necessitates that manufacturers exercise due diligence when incorporating software, implementing comprehensive risk management processes to identify and mitigate potential security vulnerabilities. Veksler et al. [43] highlight research efforts in the realm of vehicle cybersecurity, alongside two established research methodologies, namely the simulation of cyberattacks or cyberthreats on an operational test site or within a live system. The first method encompasses the formulation of cyberattack scenarios as well as the development of behavioural simulation models [21]. The latter methodology poses challenges in implementation, as only a limited number of countries have chosen to allow autonomous vehicles in public road traffic, beyond controlled test environments. For instance, Waymo and Cruise have launched autonomous taxis for public use in designated areas in the United States, including Arizona and San Francisco [3].

Furthermore, the Product Liability Directive (PLD) mandates that manufacturers systematically oversee and uphold the safety of their products throughout their entire life cycle, a task that can be particularly challenging when various components are involved. Article 4(5) of the PLD asserts that a product remains under the manufacturer's jurisdiction if the manufacturer directly governs or authorises the integration and updates of the software, even in cases where third parties are responsible for these updates. Liability persists as long as the manufacturer retains authority over the product (for instance, through updates). The newly established framework introduces the concept of no-fault liability, which is no longer limited to tangible products but will also encompass digital products and services. Software is classified as a product that is subject to the same stringent liability regulations as physical goods. This extension also applies to connected devices.

8.5.3.4 Responsible Entities

Pursuant to the Product Liability Directive, compensation claims may be initiated by a natural person who has sustained damage, by an individual who has succeeded to or been subrogated to the rights of the injured party under European Union or national law or contractual agreements, or by an individual representing one or several injured parties, as permitted by European Union or national law (Article 5(2) of the PLD). Any individual entitled to compensation is entitled to file a claim against various categories of entities.

In principle, product liability refers to the manufacturer's responsibility. The manufacturer of the product, commonly known as the OEM (Original Equipment Manufacturer), is accountable for all defects in the final product, irrespective of whether the defective component was produced internally or sourced from a supplier. This liability extends to defects arising from substandard raw materials as well as those due to unreasonable design of the product [46]. Should the product fail to meet the mandated safety standards, the manufacturer remains liable, regardless of the underlying cause of the defect. Additionally, the manufacturers of any defective components are also held responsible, provided their integration into the product was authorised by the primary manufacturer, as specified in Article 8(1) (a) (b) of the PLD. If a defective component results in damage, individuals are entitled to seek compensation from either the manufacturer of the component or the manufacturer of the product. In accordance with recital 36 of the PLD, if a manufacturer integrates a defective component from another manufacturer into the product, the injured party should have the right to claim compensation for the same damages from both the product manufacturer and the component manufacturer.

The Product Liability Directive expands the pool of liable parties to include additional defendants. These individuals are drawn from the category of intermediaries who facilitate and oversee online commerce. By incorporating them as responsible parties, their liability now extends beyond merely distributing digital products to also encompassing traditional goods. In this context, Article 7 of the PLD evaluates digitalisation from a perspective that differs from that of the product's nature: it considers digital distribution methods [46].

If the manufacturer is located outside the European Union, claims may be filed against the importer, authorised representative, or fulfilment service provider (Article 8(1)(c)(iii)). The role of the fulfilment service provider represents a novel aspect of product liability, yet it is already well established within product safety legislation. Fulfilment service providers perform numerous functions akin to those of importers, although these may not always align with the traditional definition of an importer under Union law. As economic operators, fulfilment service providers are assuming an increasingly vital role in facilitating access to the Union market for products originating from third countries. This shift in significance is duly reflected in the product safety and market surveillance framework (recital 37 of the PLD), specifically Regulation 2019/1020 [32]. In accordance with this regulation, the term 'fulfilment service provider' is defined as any natural or legal person engaging in commercial activities that offer at least two of the following services: warehousing, packaging, addressing, and dispatching, without possessing ownership of the involved products, excluding postal services as defined in point 1 of Article 2 of Directive 97/67/EC of the European Parliament and of the Council (31), parcel delivery services as delineated in point 2 of Article 2 of Regulation (EU) 2018/644 of the European Parliament and of the Council (32), and any other postal or freight transport services.

According to recital 38 of the PLD, the practice of online selling has demonstrated consistent and steady growth, leading to the emergence of new business models and market participants, including online platforms. Regulation (EU) 2022/2065 [28] and Regulation (EU) 2023/988 govern, among other aspects, the responsibility and accountability of online platforms in relation to illegal content, particularly concerning the sale of products.

When online platforms assume the roles of manufacturer, importer, authorised representative, fulfilment service provider, or distributor concerning a defective product, they ought to be subject to the same liability as these economic operators. In instances where online platforms function solely as intermediaries in the sale of products between traders and consumers, they benefit from a conditional liability exemption pursuant to Regulation (EU) 2022/2065. Nevertheless, this regulation stipulates that online platforms facilitating consumers in concluding distance contracts with traders are not exempt from liability under consumer protection law if they present the product or otherwise enable the specific transaction in a manner likely to lead an average consumer to believe that the product is provided either by the online platform itself or by a trader acting under its authority or control. In alignment with this principle, when online platforms present the product or enable the specific transaction in such a manner, it should be possible to hold them liable similarly to distributors under this directive. Consequently, the provisions of this Directive pertaining to distributors should analogously apply to these online platforms. This implies that such online platforms should be liable only when they present the product or enable the specific transaction in a manner likely to lead an average consumer to believe that the product is provided either by the online platform itself or by a trader acting under its authority or control and only in circumstances where the online platform fails to promptly identify a relevant economic operator established within the European Union.

The liability of the fulfilment service provider requires that the manufacturer of the defective product is located outside the European Union, along with the condition that neither the importer of the faulty product nor an authorised representative of the manufacturer is based within the European Union.

Consequently, this provision specifically addresses products produced outside the European Union and subsequently offered to purchasers within the European internal market, without the intermediary involvement of an importer based in the EU.

When multiple entities are liable for particular damage, they exhibit joint and several liability, which cannot be waived by contract as stipulated in Article 12(1) of the PLD. In instances where the claimant cannot identify the manufacturer or, in the case of non-EU manufacturers, any of the aforementioned entities, any distributor of the product may be held liable for the resulting damage, in accordance with Article 8(3) of the PLD. Furthermore, pursuant to Article 8(4) of the PLD, a provider of an online platform facilitating consumers in concluding distance contracts with traders, who does not qualify as an economic operator, shall also bear liability, provided that the conditions outlined in Article 6(3) of Regulation (EU) 2022/2065, also known as the Digital Services Act (DSA), are met. This provision specifies that the online platform must present the specific item of information or otherwise enable the particular transaction at issue in such a manner that would lead an average consumer to understand that the information, or the product or service involved in the transaction, is offered either by the online platform itself or by a recipient of the service acting under its authority or control.

According to Wagner [46], equating 'fulfilment service providers' with 'importers' implies that the liability of 'fulfilment service providers' takes precedence over that of 'distributors', who remain subject solely to subsidiary liability. In other words, a distributor is held liable only in situations where the manufacturer cannot be identified or is based outside the Union, and there is no involvement of an authorised representative of the manufacturer, an importer, or a fulfilment service provider domiciled within the Union (Article 8(5) of the PLD).

8.6 The Significance of Standardisation in Cybersecurity

Standardisation serves as a crucial instrument for enhancing cybersecurity. Khan et al. [22] underscore that the infrastructure of the transport system related to autonomous vehicles poses a security risk due to potential cyber-attacks, thereby necessitating rigorous cybersecurity measures and regulations. Manufacturers of vehicles equipped with embedded software, along with digital mobility service providers, fall under the NIS2 Directive. The objective is to ensure secure connections and satisfactory quality standards for the optimal functioning of autonomous vehicles. Burzio et al. [2] reference the ISO 27000 standards for Information Security Management Systems (ISMS), which align with systems advocated by other ISO standards such as ISO 9000 and ISO 14000, used for the management and control of information security risks within organisations. A fundamental principle of the ISO/IEC 27000 standards is the deliberate integration of information security under explicit management oversight. These ISO standards introduce comprehensive vocabulary, establish specific goals, provide objective-oriented validation methods, address safety concerns, and include references to vehicle cybersecurity [19]. The SAE J3061 document [37], grounded in ISO 26262, delineates risk, vulnerability, and threat nomenclature while distinguishing between system safety and system cybersecurity [39]. Furthermore, Burzio et al. [2] cite the SAE (Society of Automotive Engineers) Recommended Practice J3061 [37] for automotive cybersecurity, which offers a framework and guidelines for the cybersecurity process.

According to Colonna [5], the development of standards is critical for implementing the CRA, with a draft request for the establishment of 44 standards currently in progress [12]. Compliance with applicable harmonised standards enables manufacturers to demonstrate legal conformity. These standards are essential in delineating the specific requirements necessary for executing the CRA. The regulatory framework of the CRA is designed to oversee the primary cybersecurity requirements for interconnected devices while also allowing European Standardisation Organisations (ESOs) to further articulate these specifics. Self-assessments are permissible only if manufacturers adhere to the relevant harmonised standards.

Upon completing the required conformity assessment procedure, manufacturers must prepare an EU declaration of conformity. This document provides essential information mandated by the CRA regarding the conformity of products with digital components in relation to the fundamental cybersecurity standards established by this regulation, as well as, where appropriate, other relevant Union harmonisation legislation applicable to the product. Furthermore, manufacturers may be required to draft an EU declaration of conformity as stipulated by other Union legal acts. To improve the accessibility of information for market surveillance, a consolidated EU declaration of conformity should be compiled to demonstrate compliance with all relevant Union legal acts. To lessen the administrative burden on economic operators, this consolidated declaration may consist of a dossier containing the relevant individual declarations of conformity.

8.7 Conclusions

Defining the assumptions of legal regulations relating to cybersecurity in the context of autonomous vehicle operation necessitates consideration of global challenges posed by cybersecurity threats, alongside the specificities of Internet network functionality. This situation demands a coherent approach to addressing the issues associated with ensuring cybersecurity.

Firstly, the integration of autonomous vehicles into road traffic within the European Union shows a different level of maturity compared to what is observed in Asia and North America. In regions where the deployment of autonomous vehicles for traffic purposes has not been approved, it remains possible to gather data regarding cyber threats associated with their operation within controlled testing environments.

Secondly, the reference to the automation levels articulated by SAE [50] is made voluntarily. Consequently, there are no mandated automation levels that delineate the extent of human and Automated Driving System ('ADS') involvement in the execution of driving activities utilising an autonomous vehicle.

Thirdly, while cyberattacks often occur across national borders, the skills and policy responses of cybersecurity and law enforcement agencies remain mainly national in scope. Given the borderless nature of cyber threats, there is a pressing need to enhance capabilities at the EU level that could support the initiatives of Member States, especially considering the open borders established by the Schengen Agreement.

Fourthly, establishing a legal framework for cybersecurity particularly requires consideration of the technical aspects of data processing via autonomous vehicles and road infrastructure components. Additionally, this legal framework should incorporate standards for intentional human behaviour to protect data from cybersecurity threats.

Khan et al. [22] outlined the assumptions for a regulatory framework designated as the 'cybersecurity regulatory framework', referred to by the authors as the 'CRF'. This framework is grounded in the following principles:

(1) Liability—denoting the state of being legally accountable for the operation of autonomous vehicles;
(2) Privacy—the fundamental autonomy of consumers, along with the protection of the intellectual property of automakers;
(3) Ethics and Moral Decision-Making—the capacity of autonomous vehicles to learn from and evaluate real-world situations, such as interactions with law enforcement and emergency services;
(4) Accessibility—addressing the distinct needs, expectations, and desires of Connected and Automated Vehicle (CAV) users; and
(5) Equity—considering the spatial and temporal fairness of autonomous operations.

The initiative to establish a distinct legal framework for promoting 'cybersecurity' in relation to the use of autonomous vehicles represents a prudent solution. Based on the aforementioned principles, it can be concluded that the scope of the CRF legal framework under consideration is extensive. This approach necessitates evaluating whether such regulations will intersect with the legal norms of the applicable legal order concerning cybersecurity. Inquiring about this matter involves incorporating this concept into the relevant legal sources, thereby transcending its abstract approach. Presuming that existing regulations related to cybersecurity do not comprehensively address the issues surrounding the use of autonomous vehicles, they should not be disregarded. It is advisable to reference relevant regulations, including the NIS2 Directive and the Cybersecurity Act.

Consequently, the extent of the application of the existing legal framework must be established, for instance, by consulting the relevant regulations. This necessitates determining whether the definitional framework established by the Cybersecurity Act applies to an autonomous vehicle and, most importantly, whether an autonomous vehicle qualifies as an ICT product under the definition provided in Article 2(12) of this Regulation.

The primary focus of the discussion regarding the establishment of a legal framework for cybersecurity relates to the operation of autonomous vehicles, which encompass the varying levels of automation required for autonomous driving (SAE 2018). These vehicles act both as a source and a contributor to the accumulation of substantial volumes of data, which, due to the use of a radio network for data collection and subsequent processing, is vulnerable to cyberattacks.

An autonomous vehicle serves as a point of reference for delineating the various entities involved in: (1) the creation and production process, as well as the development of digital mobility services that influence the vehicle's driving functionality; and (2) the approval process for its use in traffic, including the inspection and oversight of the road infrastructure that facilitates the autonomous vehicle's operation in the operational driving mode (ODD) and the vehicles themselves. In this manner, the necessary data flow mandated to perform the responsibilities pertinent to each stakeholder, aimed at securing the vehicle's approval for road use, can be established. The establishment of this data flow (encompassing personal data) would enable the identification of the entities accountable for its security within the framework of the specific responsibilities designated to the roles these entities fulfil.

Wang et al. [47] elucidates the prerequisites for applying artificial intelligence concerning significant issues, including algorithmic bias, privacy infringements, and threats to human rights and democratic values. As a result, the global governance of artificial intelligence has emerged as a crucial policy domain, with governments adopting various strategies to balance innovation with regulatory measures. For instance, Regulation EU 2024/1689 ('AI Act') [34] employs a risk-based framework, categorising AI systems based on their potential impact on human rights and safety, and prohibiting those classified as unacceptable, such as real-time biometric identification.

References

1. Brüggemeier, G.: Tort Law in the European Union. Kluwer Law International (2015) 177; Verbruggen and Others (n 37) 99–100
2. Burzio, G., Cordella, G.F., Colajanni, M., Marchetti, M., Stabili, D.: Cybersecurity of connected autonomous vehicles: a ranking based approach. In: 2018 International Conference of Electrical and Electronic Technologies for Automotive, pp. 1–6. IEEE (2018)
3. Chari, M.: Points of reference: robotaxi safety. https://www.apmresearchlab.org/10x/robotaxi-safety
4. Cloudflare: https://www.cloudflare.com/learning/cdn/glossary/internet-exchange-point-ixp/ (2025). Accessed 12 Feb 2025
5. Colonna, L.: The end of open source? Regulating open source under the cyber resilience act and the new product liability directive. Comput. Law Secur. Rev. **56**, 106105 (2025). https://doi.org/10.1016/j.clsr.2024.106105
6. Council Directive 85/374/EEC of 25 July 1985 on the Approximation of the Laws, Regulations and Administrative Provisions of the Member States Concerning Liability for Defective Products, OJ L 210, 7.8.1985, pp. 29–33
7. Directive (EU) 2016/1148 of the European Parliament and of the Council of 6 July 2016 Concerning Measures for a High Common Level of Security of Network and Information Systems Across the Union OJ L 194, 19.7.2016, pp. 1–30
8. Directive (EU) 2018/1972 of the European Parliament and of the Council of 11 December 2018 Establishing the European Electronic Communications Code (Recast) (Text with EEA Relevance), PE/52/2018/REV/1, OJ L 321, 17.12.2018, pp. 36–214
9. Directive (EU) 2022/2555 of the European Parliament and of the Council of 14 December 2022 on Measures for a High Common Level of Cybersecurity Across the Union, Amending Regulation (EU) No 910/2014 and Directive (EU) 2018/1972, and Repealing Directive (EU) 2016/1148 (NIS 2 Directive) (Text with EEA Relevance), PE/32/2022/REV/2, OJ L 333, 27.12.2022, pp. 80–152
10. Directive (EU) 2022/2557 of the European Parliament and of the Council of 14 December 2022 on the Resilience of Critical Entities and Repealing Council Directive 2008/114/EC (Text with EEA Relevance), PE/51/2022/REV/1, OJ L 333, 27.12.2022
11. Directive (EU) 2024/2853 of the European Parliament and of the Council of 23 October 2024 on Liability for Defective Products and Repealing Council Directive 85/374/EEC (Text with EEA Relevance), PE/7/2024/REV/1, OJ L, 2024/2853, 18.11.2024, ELI
12. Draft Standardization Request to European Standards Organizations in Support of Union Policy on Cybersecurity Requirements for Products with Digital Elements, 17 Apr 2024. https://ec.europa.eu/docsroom/documents/58974. Accessed 1 Feb 2025
13. Dukarski, J.: Unsettled legal issues facing data in autonomous, connected, electric, and shared vehicles. SAE Research Report EPR2021019 (2021). https://doi.org/10.4271/EPR2021019
14. European Commission: Impact Assessment Report Accompanying the Document Proposal for a Directive of the European Parliament and of the Council on Measures for a High Common Level of Cybersecurity Across the Union, Repealing Directive (EU) 2016/1148—Part 1, SWD (2020) 345 Final, 16
15. European Commission: Impact Assessment Report Accompanying the Document Proposal for a Directive of the European Parliament and of the Council on Measures for a High Common Level of Cybersecurity Across the Union, Repealing Directive (EU) 2016/1148—Part 2, SWD (2020) 345 Final, 15
16. Eurostat: 22% of EU enterprises had ICT security incidents. Eurostat, 14 Feb 2023. https://ec.europa.eu/eurostat/web/products-eurostat-news/w/edn-20230214-1#:~:text=In%202021%20in%20the%20EU,of%20ICT%20services%2C%20destruction%20or. Accessed 4 Jan 2025
17. IBM Security: Cost of a Data Breach Report 2022 (2022). ibm.com. https://www.ibm.com/reports/data-breach. Accessed 5 Jan 2025

18. International Telecommunications Union (ITU): ITU-TX.1205: series X: data networks, open system communications and security: telecommunication security: overview of cybersecurity 2008. https://www.itu.int/rec/T-REC-X.1205-200804-I/en. Accessed 14 Feb 2025
19. ISO road vehicles functional safety. Part_1: standard ISO 26262-1. https://www.iso.org/standard/68383.html (2018). Accessed 22 July 2022. Accessed 21 Apr 2024
20. Judgment of the Court (First Chamber) of 10 June 2021, Case C-65/20, VI v KRONE—Verlag Gesellschaft mbH & Co KG (ECLI:EU:C:2021:471)
21. Kavak, H., Padilla, J.J., Vernon-Bido, D., Diallo, S.Y., Gore, R., Shetty, S.: Simulation for cybersecurity: state of the art and future directions. J. Cybersecur. **7**(1), tyab005 (2021)
22. Khan, S.K., Shiwakoti, N., Stasinopoulos, P., Warren, M.: Cybersecurity regulatory challenges for connected and automated vehicles—state-of-the-art and future directions. Transp. Policy **143**, 58–71 (2023). https://doi.org/10.1016/j.tranpol.2023.09.001
23. Khan, S.K., Shiwakoti, N., Stasinopoulos, P., Chen, Y., Warren, M.: Exploratory factor analysis for cybersecurity regulation and consumer data in autonomous vehicle acceptance: insights from four OECD countries. Transp. Res. Interdiscip. Perspect. **25**, 101084 (2024)
24. Khan, S.K., Shiwakoti, N., Stasinopoulos, P., Chen, Y., Warren, M.: Cybersecurity framework for connected and automated vehicles: a modeling perspective. Transp. Policy **162**, 47–64 (2025)
25. Koch, B.A.: Product liability on the way to the digital age. J. Eur. Tort Law **15**, 109–125, 117 (2024)
26. Miller, M.R., Herrera, F., Jun, H., Landay, J.A., Bailenson, J.N.: Personal identifiability of user tracking data during observation of 360-degree VR video. Sci. Rep. **10**(1), 1–10 (2020)
27. Morgan, S.: Cybercrime to cost the world $10.5 trillion annually by 2025. Cybercrime Magazine, 13 Nov 2020. https://cybersecurityventures.com/cybercrime-damage-costs-10-trillion-by-2025. Accessed 5 Jan 2025
28. Regulation (EU) 2022/2065 of the European Parliament and of the Council of 19 October 2022 on a Single Market For Digital Services and amending Directive 2000/31/EC (Digital Services Act) (Text with EEA relevance), PE/30/2022/REV/1, OJ L 277, 27.10.2022, p. 1–102.
29. Raiyn, J.: Data and cyber security in autonomous vehicle networks. Transp. Telecommun. **19**(4), 325–334 (2018)
30. Regulation (EU) 2016/679 of the European Parliament and of the Council of 27 April 2016 on the Protection of Natural Persons with Regard to the Processing of Personal Data and on the Free Movement of Such Data, and Repealing Directive 95/46/EC, OJ L 119, 4.5.2016, pp. 1–88
31. Regulation (EU) 2019/881 of the European Parliament and of the Council of 17 April 2019 on ENISA (The European Union Agency for Cybersecurity) and on Information and Communications Technology Cybersecurity Certification and Repealing Regulation (EU) No 526/2013 (Cybersecurity Act) (Text with EEA Relevance), PE/86/2018/REV/1, OJ L 151, 7.6.2019, pp. 15–69
32. Regulation (EU) 2019/1020 of the European Parliament and of the Council of 20 June 2019 on Market Surveillance and Compliance of Products and Amending Directive 2004/42/EC and Regulations (EC) No 765/2008 and (EU) No 305/2011 (Text with EEA Relevance), PE/45/2019/REV/1, OJ L 169, 25.6.2019, pp. 1–44
33. Regulation (EU) 2022/2554 of the European Parliament and of the Council of 14 December 2022 on Digital Operational Resilience for the Financial Sector and Amending Regulations (EC) No 1060/2009, (EU) No 648/2012, (EU) No 600/2014, (EU) No 909/2014 and (EU) 2016/1011 (Text with EEA Relevance), PE/41/2022/INIT
34. Regulation (EU) 2024/1689 of the European Parliament and of the Council of 13 June 2024 Laying Down Harmonised Rules on Artificial Intelligence and Amending Regulations (EC) No 300/2008, (EU) No 167/2013, (EU) No 168/2013, (EU) 2018/858, (EU) 2018/1139 and (EU) 2019/2144 and Directives 2014/90/EU, (EU) 2016/797 and (EU) 2020/1828 (Artificial Intelligence Act) (Text with EEA Relevance), E/24/2024/REV/1, OJ L, 2024/1689, 12.7.2024, ELI: http://data.europa.eu/eli/reg/2024/1689/oj. Accessed 21 Feb 2025
35. Regulation (EU) 2024/2847 of the European Parliament and of the Council of 23 October 2024 on Horizontal Cybersecurity Requirements for Products with Digital Elements and

Amending Regulations (EU) No 168/2013 and (EU) 2019/1020 and Directive (EU) 2020/1828 (Cyber Resilience Act) (Text with EEA Relevance), PE/100/2023/REV/1, OJ L, 2024/2847, 20.11.2024
36. Regulation (EU) No 526/2013 of the European Parliament and of the Council of 21 May 2013 Concerning the European Union Agency for Network and Information Security (ENISA) and Repealing Regulation (EC) No 460/2004 (OJ L 165, 18.6.2013, p. 41)
37. S. V. E. S. S. Committee, SAE J3061-Cybersecurity Guidebook for Cyber-Physical Automotive Systems. SAE-Society of Automotive Engineers (2016)
38. Schatz, D., Bashroush, R., Wall, J.: Towards a more representative definition of cyber security. J. Digit. Forensics Secur. Law **12**(2), 8 (2017)
39. Sun, X., Yu, F.R., Zhang, P.: A survey on cyber-security of connected and autonomous vehicles (CAVs). IEEE Trans. Intell. Transp. Syst. (2021)
40. Tasheva, I.: Cybersecurity post-COVID-19: lessons learned and policy recommendations. Eur. View **20**, 140–141 (2021)
41. Tettenborn, A.: In: Jones, M.A. (ed) Clerk & Lindsell on Torts, 23rd edn., ch 10, para 52 (2020)
42. Vandezande, N.: Cybersecurity in the EU: how the NIS2-directive stacks up against its predecessor. Comput. Law Secur. Rev. **52**, 105890 (2024). https://doi.org/10.1016/j.clsr.2023.105890
43. Veksler, V.D., Buchler, N., Hoffman, B.E., Cassenti, D.N., Sample, C., Sugrim, S.: Simulations in cyber-security: a review of cognitive modeling of network attackers, defenders, and users. Front Psychol. **9**, 691 (2018)
44. Vellinga, N.E.: Cyber security in (automated) vehicles and liability: the EU legal framework and (a lack of) compensation. Transp. Res. Procedia **72**, 132–138 (2023)
45. Verizon: 2021 Data Breach Investigations Report. Verizon.com (2021). https://www.verizon.com/business/resources/reports/2021-data-breach-investigations-report.pdfx. Accessed 2 May 2025
46. Wagner, G.: Liability rules for the digital age—aiming for the Brussels effect. J. Eur. Tort Law (2023). https://doi.org/10.1515/jetl-2022-0012
47. Wang, S., Zhang, Y., Xiao, Y., Liang, Z.: Artificial intelligence policy frameworks in China, the European Union and the United States: an analysis based on structure topic model. Technol. Forecast. Soc. Change **212**, 123971 (2025). https://doi.org/10.1016/j.techfore.2025.123971
48. Wolters, P.T.J.: The obligation to update insecure software in the light of Consumentenbond/Samsung. Comput. Law Secur. Rev. **35**(3), 295–305 (2019)
49. Draft Standardisation Request to European Standards Organisations in Support of Union Policy on Cybersecurity Requirements for Products with Digital Elements (17 April 2024); https://www.ec.europa.eu/docsroom/documents/58974, accessed: 1 Feb 2025.
50. SAE: Taxonomy and Definitions for Terms Related to Driving Automation Systems for On-Road Motor Vehicles J3016_202104, (2021) https://www.eproofing.springer.com/ePb/books/-3S6Nde8RWQpK4wf100r5_BGRE1YO3bEDkNBb_JWyyety1oQvwfb_2dJNrMuikQQQouawjvX_hyawEu5gIBiUuvlPThIysvY1iMv3ckWALO40Xi5n9htb6Jwoj__P2pSgWqBN-qpCJBdVm1IJer8dg==/authseq=1, Accessed 1 Feb 2025.

Summary

The evolution of business models from a product-centric approach to a service-centric one is driven by the need to generate market value. As a fundamental business objective, the entrepreneur's value creation through the business model relates to strategies for producing, delivering, and capturing value. This evolution particularly emphasises critical components such as infrastructure, business offerings, customer base, and financial framework. Value creation can be evaluated from both the entrepreneur's perspective, as a supplier of products and services, and the customer's perspective, as the beneficiary, indicating mutual advantages.

In the realm of transportation, several relationships among entities can be discerned: (1) the user is the proprietor of the product, and the provider augments value by offering guidance on its effective utilisation; (2) the customer utilises the service, while the provider bears responsibility for the quality of the service (e.g., selecting an appropriate route, driver, or type of vehicle); (3) the customer compensates the mobile service; however, the service provider ascertains the optimal transportation option (e.g., route plan) and the most suitable infrastructure option (e.g., road vehicle). The transition from products also involves acquiring solutions through the development of services. This progression advances from less sophisticated services to more advanced services, ultimately culminating in solutions along the product-service continuum. Empirically, this transition is often referred to as 'service infusion'. These considerations necessitate a differentiation between the concept of 'servitisation', which should be understood within a theoretical framework, as an empirical phenomenon, and as a domain of research. It pertains to the relationship between the manufacturer and the customer, necessitating intra-organisational changes. These changes incorporate the mission, vision, and business strategy of the manufacturer, specifically, the transformation of the business model from one centred on product provision to a model that exclusively offers services or provides services alongside products or other functionalities, based on customer requirements. The degree of changes introduced in the entrepreneur's organisational realm, along

A. Krasuski, *Legal Challenges of a Transformation from a Product-Centric to a Service-Centric Business Model and Procedures*, Studies in Systems, Decision and Control 610, https://doi.org/10.1007/978-3-031-98901-8

with the management of resources, is justified by the transition in the business model from a product-oriented to a service-oriented framework.

The use of autonomous vehicles, along with the accessibility of digital tools, acts as a significant catalyst for transforming business models within the transportation sector. This shift towards service-oriented business models is justified, as vehicles—considered as products—integrate services that create value for both manufacturers and associated partners. Certain products, including vehicles, inherently encompass services. In redefining business models, entrepreneurs must also consider the availability of digital tools that enhance customer communication and facilitate data analysis to support service delivery. Functionally, the term 'digitalisation' relates to the instruments used for the exchange of information among entities and objects. In this context, the information exchange among objects is intended to assist entities.

Servitisation necessitates strategic decisions at the highest levels of management. Such decisions can significantly impact the development of customer relationships and, consequently, value creation, thereby contributing to revenue generation and entrepreneurial growth. Legal regulations constitute a crucial factor influencing the feasibility of transitioning from a product-oriented to a service-oriented business model. Thus, they must be considered in the process of shaping business models. The legal permissibility of conducting business activities as expressed in business models is essential for their implementation, alongside non-legal factors such as access to appropriate resources, necessary organisational changes, and demand for specific digital mobility services. The law encompasses risks that affect the efficacy of implementing the business model, and these risks can be multifaceted. Firstly, it is imperative to evaluate how the principle of economic freedom is upheld in the countries where economic activities involving the provision of digital mobility services are conducted. Economic freedom should not be misconstrued as arbitrary; rather, it should entail proportionate legal requirements as conditions for initiating and conducting business activities. The establishment of state or state-related monopolies, combined with legal requirements that are disconnected from market realities and practically unfeasible to implement, culminates in legal barriers that hinder this type of business activity. The aforementioned assertions are substantiated by the first thesis cited in the monograph. Legal barriers may induce initial stakeholder decisions against the implementation of business models, either due to prohibitions on conducting business activities or due to an existing formalised regulatory framework related to the business model that encumbers business operations with excessive risk and, consequently, costs. Such barriers may also materialise after the initiation of business activities as part of the business model implementation, potentially leading to a decision to 'deservitise', which arises from the failure of the employed business models. This term encompasses two scenarios. On one hand, 'deservitise' refers to a reversion to 'product-centric' business models. The second option involves the complete cessation of business activity.

The complexity of situations to which legal obligations pertain presents a significant challenge for entrepreneurs considering the implementation of digital mobility services. Regulation may also act as a barrier to experimentation; for example, it may obstruct efforts to accelerate the transition to a low-carbon economy or to tackle liability issues associated with autonomous driving. Disruptive innovations can trigger important public policy concerns, necessitating new forms of regulation. It is crucial to recognise that establishing a legal framework for providing digital mobility services poses a considerable challenge for legislators. The legal landscape governing road transport automation, including product liability, cybersecurity, and intellectual property, is of great importance. Furthermore, the regulatory framework for implementing digital mobility services should include the oversight responsibilities of regulatory bodies. Notably, digital mobility services are provided by companies in collaboration with third parties within the established ecosystem. A key feature of the business models related to the implementation of digital mobility services is the shift of service offerings and purchasing methods to the internet, as services available over-the-top. A prerequisite for becoming familiar with the available services and acquiring these offerings is the customer's access to the internet, which is a technical requirement. Nevertheless, the marketing of digital mobility services, as well as their delivery through an intermediary entity, such as a digital platform operator, is still subject to legal regulations governing e-commerce. Digital mobility services are realised in conjunction with business models devised through collaboration between vehicle manufacturers and technology firms. Digitalisation emerges as a central factor in the advancement of these business models. It plays a crucial role in influencing shifts in customer preferences for mobility services, which can be obtained online via specialised applications. Moreover, digitalisation supports the development of research and commercial applications for autonomous vehicles, whether they operate independently or as part of a connected network of semi-autonomous and fully autonomous vehicles. The potential use of autonomous vehicles alongside digital mobility services creates opportunities to customise a wide range of these services to meet market demands. The vast opportunities presented by digital mobility services are based on the concept of 'X-as-a-service' as articulated in relevant literature.

The term 'digital mobility service' does not have a legal definition; its significance has been articulated based on the defined business models of digital mobility. By referencing these models, one can categorise this service into five distinct categories: (1) access to transport services provided by a third party using its vehicle, (2) access to a vehicle for travel, (3) assistance for the driver with tasks directly related to planning and executing trips, (4) access to content (including voice and data) that does not pertain to trip planning and execution but rather enhances the travel experience, and (5) support in managing passenger and freight transport. These services are delivered over the internet by providers operating within an ecosystem that often includes the vehicle manufacturer. Such services are provided for compensation and are considered voluntary offerings associated with autonomous wheeled transport.

The legislation also influences how digital mobility services are delivered, thereby affirming the proposed second research thesis. The impact of the legal framework on the provision of digital mobility services is multifaceted. It should be considered in relation to both the subject of digital mobility services and their mode of delivery. The law does not define digital mobility services. Furthermore, there is a notable lack of dedicated legal regulations concerning the provision of such services. It is imperative to introduce these regulations at the EU level. Under the current legal framework, identifying the appropriate legal sources must take into account the nature of these services, the demographics of recipients, and the method of their delivery. The subject matter of digital mobility services should be delineated based on the identified categories of these services. These services are facilitated within an ecosystem that encompasses the collaboration of various entities. The fundamental legal relationship regarding the provision of these services involves the service provider, the recipient, and the entity that intermediates the conclusion of contracts through digital platforms. The significant market position of digital platforms may lead to distortions of competition, a concern acknowledged by the EU legislator. Within the framework of EU law, the legislator has modified the legal structure to mitigate the market dominance of digital platforms, including through the implementation of the Digital Markets Act (DMA). The enforcement of EU competition law provisions, including established competition law precedents, does not undermine the application of the DMA; conversely, it does not impede it. Based on this EU regulation, the European Commission has acquired authority and has designated a total of 24 gatekeepers that provide essential platform services. The DMA imposes a series of obligations on these gatekeepers, including, among others, providers of online intermediation services, many of which directly pertain to online intermediation services. On one hand, it is essential to recognise the advantages of EU regulations aimed at the operations of digital platforms, manifested in the increased diversification of the conditions under which products and services are offered. This aims to enhance price competition and benefit end users while also fostering competition in the realm of online intermediation services rendered by the gatekeepers. On the other hand, however, these regulations complicate cooperation with digital platform operators due to stringent regulatory mandates.

In the context of providing digital mobility services, the law delineates the legal status and scope of rights for individuals who serve as recipients of these services, as well as the status of vehicle drivers, including those of autonomous vehicles, in relation to them. The legislator has conferred upon consumers certain rights that are protected under both EU law and the national legislation of the Member States. Nevertheless, there is a lack of comprehensive regulations addressing the operation of autonomous vehicles and the legal liabilities—civil, criminal, and administrative—arising from incidents related to their movement. Accordingly, two principal issues warrant consideration. Firstly, it is essential to initiate a discourse regarding the parameters and preparedness necessary to grant a distinct legal status to artificial agents (ADS) concerning autonomous vehicles operating in high and full automation modes. As long as a human driver is regarded as the recipient of digital mobility services, providing such services is inherently constrained by the human capacity to process information in a way that does not jeopardise road safety. In the context of employing autonomous vehicles, designating a human being as the recipient of digital mobility services restricts the provision of these services to the third level of automation, as classified by SAE. The second issue involves the regulation of the algorithms utilised in the learning processes based on data acquired from the driving environment. Drawing from this learned experience, the algorithm formulates its internal decision-making logic to optimise task performance. Once this logic is established, the algorithm can autonomously make decisions regarding driving in dynamic environments, independent of human intervention. It is impractical to comprehensively test vehicles against an indefinite array of unforeseen circumstances that may arise in the driving environment. In extreme situations necessitating rapid responses in road traffic, decisions executed by ADS are intentional, even in scenarios that entail the "sacrifice" of one good for another. The foundation of these decisions resides within the algorithms, particularly in the manner in which they are programmed. This approach may result in undue favouritism towards certain road users, thereby influencing the risk associated with the autonomous vehicle's behaviour on the road. The rationale for this preferential treatment is grounded in the moral implications of the decisions made. The legislator ought to analyse and subsequently regulate the mechanisms governing the method of risk distribution, delineating what is morally permissible concerning the decisions made by ADS pertaining to the operation of autonomous vehicles. In this context, the legislator should establish regulations governing the criteria for decision-making influenced by algorithms.

The provision of digital mobility services entails processing substantial volumes of data that facilitate the identification of natural persons, thereby constituting personal data. As users' data from digital mobility services is also processed alongside the provision of internet access services, it is imperative to consider the data protection requirements encompassed by the confidentiality of electronic communications. The legal protection of personal data derives from safeguarding specific values, serving as the foundation for establishing a legal framework in this domain. In the context of EU law, the source of legal protection for personal data is the principle of human dignity. In national legislation across Member States, alongside human dignity, other values are recognised as sources for data protection regulations, including, for instance, informational autonomy. The intricacy of the regulations governing the protection of personal data is manifested in the multifaceted nature of legal requirements, which encompass: possessing a legal basis for processing each category of personal data and its associated purpose; in the absence of such a basis, the data must be deleted or anonymised; safeguarding personal data, ensuring its security and availability solely based on existing legal justifications; fulfilling obligations towards individuals whose data are processed in accordance with their rights under the General Data Protection Regulation (GDPR); and adhering to the stipulations set forth by the national data protection authority in alignment with legal provisions. Before initiating data processing, the data controller must implement a multi-step approach: first, evaluating how the proposed processing activities will influence the protection of personal data. The GDPR may not be deemed entirely effective regarding digital platforms in safeguarding the rights and freedoms of individuals due to several factors: the extensive collection of personal data from users, the complex nature of service provision, the resulting intricate data flows, and the implications for data subjects, other entities within the digital ecosystem, compliance with GDPR obligations, and the restricted capacity to verify and enforce these obligations.

The challenge related to data processing in the context of providing digital mobility services, particularly concerning the use of autonomous vehicles, is the necessity of ensuring the cybersecurity of such data. Defining the legal regulatory assumptions regarding cybersecurity within the realm of autonomous vehicle traffic necessitates consideration of the global challenges associated with cybersecurity threats, as well as the specific operational characteristics of the Internet, which demand a coherent approach to ensuring cybersecurity. Given the borderless nature of cyber threats, there exists a pressing need to enhance capabilities at the EU level that can complement the actions undertaken by Member States, particularly due to the open borders established within the Schengen area. Furthermore, the establishment of a legal framework for cybersecurity must specifically consider the technical aspects of data processing through autonomous vehicles and the elements of road infrastructure. Additionally, this legal framework should incorporate standards of informed human behaviour to safeguard data against cybersecurity threats.

The manufacturer's authorised representative in the EU is Springer Nature Customer Service Centre GmbH, Europaplatz 3, 69115 Heidelberg, Germany. If you have any concerns regarding our products, please contact ProductSafety@springernature.com

Printed and bound by CPI Group (UK) Ltd, Croydon, CR0 4YY
30/09/2026
02225570-0004